The Human Enigma

L. Mason Jones served a number of years in the military, and travelling on a so-called 'government service' passport, found himself in such places as south Yemen, Bahrain, the Gulf of Oman, Cyprus and Germany. After leaving the service, he became part of the team producing the highly successful business jet, The Hawker 125. He functioned as a quality-engineering inspector with, initially, British Aerospace then Corporate Jets Inc and finally Raytheon USA, the latter of which purchased the thriving business and moved production to the USA. Mr Jones then left the business to concentrate on writing projects. He has three adult offspring and resides in Chester.

By the same Author –

Monkey Trial 2000
Pillars of Fire
When The Moon Came
Cultural Shock

The Human Enigma

L. Mason Jones

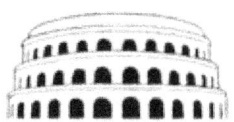

Arena Books

Copyright © L. Mason Jones, 2021

The right of L. Mason Jones to be identified as author of this book has been asserted in accordance with the Copyright, Designs and Patents Act 1988. All characters and events described in this book are fictional and any resemblance to actual persons, living or dead, is purely coincidental.

First published in 2021 by Arena Books

Arena Books
6 Southgate Green
Bury St. Edmunds
IP33 2BL

www.arenabooks.co.uk

Distributed in America by Ingram International, One Ingram Blvd., P.O. Box 3006, La Vergne, TN 37086-1985, USA.

All rights reserved. Except for the quotation of short passages for the purposes of criticism and review, no part of this publication may be reproduced, stored in a retrieval system, or transmitted, in any form or by any means, electronic, mechanical, photocopying, recording or otherwise, without the prior permission of the author or the publisher acting as his agent.

L. Mason Jones
The Human Enigma

British Library cataloguing in Publication Data. A Catalogue record for this book is available from the British Library.

ISBN-13 978-1-911593-90-4

BIC classifications:- PDA, PGK, RBC, RBGD, PDZ, ABGF, PGS, RNR, TTD.

Cover design
by Jason Anscomb

Typeset in
Times New Roman

I would like to dedicate this book to all those people who, by their efforts, are attempting to resolve the question of human origins, no matter how the answer may astound us.

Author

The trust must be found but woe to him who finds it

Robert Charroux

THE UNITED NATIONS ECONOMIC AND SOCIAL COUNCIL, (UNESCO), SPONSORED A CONFERENCE IN PARIS. ANTHROPOLOGISTS, ARCHAEOLOGISTS ETC., FROM THIRTY-FIVE DIFFERENT COUNTRIES WERE GIVEN THE PROBLEM:

"WHO IS MAN AND WHERE DID HE COME FROM"?

THE CONCLUSION OF THE SYMPOSIUM WAS:

"THE ONLY THING WE CAN BE CERTAIN OF, IS THAT WE ARE UNCERTAIN"

SYNOPSIS OF THE POETIC KIND

Humans will certainly settle on Mars, but what then? We ask....

A trip to the stars?

What if we get there and find we are alone?

Will the truth of our origins *ever* be known?

We know where we are going, but not where we are from,

Will some, 'revelation' reveal our real home?

Can *we* thank a primate for such a fine brain?

That compels us to leave our earthly domain?

Science assured us, but proof is so sparse, our forebears *were* primates and not 'Men from Mars'

But many a writer has said, "It's for sure" that E.T. indeed once came to our shores.

But did a great being, as Genesis tells, give us our bodies, our bones and our cells?

Did a primate bequeath then, this strange brain of ours, that drives us to study, then leave for the stars?

We use but a third of its cellular mass, but great men of science few can surpass.

Huge bone constructions from so long ago, adorn our museums *their* history is known, but as for the human?.........

There's hardly a bone.

But is it all part of a much greater plan, to see other worlds and enhance life thereon? Then just like our forebears, man in his turn will urge other creatures to leave *their* domain.

Our bodily atoms all came from a star and one day we'll find out who *we* really are.

CONTENTS

SYNOPSIS OF THE POETIC KIND

Chapter 1	**Missing Links**	PAGE 1
Chapter II	**The Real Ancestors**	PAGE 69
Chapter III	**God and Lesser Gods**	PAGE 77
Chapter IV	**The Cosmic Connection**	PAGE 91
EPILOGUE	**Towards The Infinite**	PAGE 151
REFERENCES		PAGE 155

Chapter I

MISSING LINKS

This work is primarily concerned with the unresolved problems of human origins. During the long and patient search that has so frustrated the paleoanthropologists in their quest for their 'missing link', or so called Pro-Consul, that has occupied their time for so many years, it has been necessary to have two strong points on which all subsequent assumptions, finds and analysis are firmly based.

These anchor points are, accurate and reliable dating processes, and the correct identification of the creature assumed to have possessed the ancient fragments in the first place. A good analogy of the importance of these anchor points would be an amusing little story and a couple of quotation, one from the British Museum press and one from the result of dating anomalies during the Moon rock analysis of the Apollo programme. From the age of steam there is a story concerning an exuberant, but rather self-important little chap, called 'Fred the Wheel tapper. One day, he brought the entire regional railway network to a standstill because, when he tapped the wheels of all the trains instead of a nice reverberant ring, all he got was a dull thud. Later, when everyone had missed their train it was discovered that it was his *hammer* that was cracked.

Not all scientists are absolutely convinced that our current methods of dating organic material found under ancient rocks are absolutely accurate and infallible, and scientific journals often warn paleoanthropologists of the perils of making wide sweeping assumptions regarding their bone fossil finds. Headlines have often declared 'missing links found', only to be refuted later by others within their science.

It can be seen therefore that there is a great potential for error. This is not to say that anthropology is not aware of this. However, in the early days of the search to prove this new and dynamic theory as soon as possible, the risk of error must have been more prevalent. One might go as far as to say that bone fossils largely skulls may have found their way on to the fossil table that possibly should not be there.

Although Charles Darwin was not over pedantic regarding his theory (and initially of Alfred Russell Wallace), he seemed quite convinced of its validity but was gracious enough to admit the proof must be found. As he appeared to be rather fixated on chimpanzees and gorillas as distant relatives of the human, his followers and those who wanted to believe in this dynamic new theory that appealed to their wish to find an adequate substitute for what many of the more atheistic among them referred to rather irreverently as a fairy tale. I refer of course to divine creation of the Genesis kind. As science, advanced into the next century a more scientific approach became necessary to eliminate exuberance and excitement whereas earlier practically every skull portion was proclaimed the 'missing link' that must of course be found.

Whereas Charles Darwin could settle back a little and hope his followers would succeed in the search, he had put a great pressure on them so much so they purposely carried out deliberate fraud in the early 1900's indicating of course the aforesaid pressure to resolve the issue by 'other means'. When more advanced dating processes emerged that discovered this monumental deception, the question must have been asked, "Was this system applied to the many bones and skulls that had already established themselves as ancestral fragments and found their place on the fossil table?"

When we look at some of the books and journals compiled during this ever lengthening search, it was clear that the initial exuberance was giving way to a more rational and analytical approach to the subject.

One of the best works dealing with the subject was *The Origin of Human Kind* (phoenix 1994) by Richard Leaky. It is clear in his statements indicating rationality and making it clear that their discoveries of the old bones must tread lightly'. It is also clear, he is (one might say) 'compelled but not convinced' regarding the theory.

He pointed out that a British Scientist had closely analysed Darwin's work and the statement therein, that seemed more towards encouraging a belief in the theory, rather than outlining factors of a compelling nature. These remarks were in the subjunctive mood and there were hundreds of them. Remarks such as "We may well suppose" and "Let us assume" etc.

It is obvious that Richard Leaky wished to 'rein in' those ardent believers in the theory both amateur and professional, who excitedly held up every old bone as a candidate for the missing link. Richard highlighted just how breathtaking Darwin's statement was, when he proclaimed that chimpanzees and gorillas were mans closet living relatives, when hardly (if any) ancient fossils that seemed to indicate this, where found. Richard made it clear that his statement was far from factual but entirely theory. He made it clear that the important issue of human origins remains entirely unresolved.

When one considers just how sparse the human fossil table really is, it must be asked 'why is this?' This paucity of convincing human ancient ancestral fossils, is made even more mysterious, when we recollect on the abundance of fossilised bones from sixty five million years ago of creatures who (it is established) met their demise due to some great cosmic conflagration at that time. The aforementioned ardent, one might almost say, 'frantic' search for proof of the Darwinian theory in the quest for the vital links, turned up many of the ancient dinosaur bones, so many, in fact that complete assemblies have been erected and can be viewed in our natural history museums. Why are they so near the surface, when, from such a huge amount of time, layer after layer of all the creatures including eventually humans expired after they were laid down? Of course, over many millennia since their demise, major geological disturbances and upheaval have occurred, but this is not an adequate explanation for all of them. However, let us consider that if the human ancestral bones had all been found, then we could construct our history as easy as we can do with the dinosaur story with all the skeletons depicting the changes over the millennia from Simian kind to humans.

However, this work is not primarily concerned regarding only the Darwinian Theory for how humans 'came to be'. It intends to look at and analyze all the theories and assumptions people hold with regard to their beginnings.

Although to a great extent, the dynamic new theory introduced by Charles Darwin disturbed the religious hierarchy and indeed many who were clinging to their faith and had endured many disparaging remarks regarding their beliefs as a mystical unscientific fairy tale, and quite unbelievable many kept their faith in the knowledge that Darwin's version was still unconfirmed.

Also, it is clear that so many books appeared with such titles as 'The Search for Jesus', did Jesus exist? Jesus, the man that the search for proof and reassurance of their faith and beliefs was sought equally as ardently as the 'apes to humans' concept.

For many people in the late 1800's Darwin's theory was a great revelation, certainly to the atheists and also to those wavering and rather lukewarm with regard to their beliefs. Many flocked to the banner of this man who could in a sense be termed 'the new Messiah'.

Many people will now of course, be aware, that the straight fight that it used to be and indeed still is, in certain Southern States of the USA, has now, due to the appearance of popular culture, recently been joined by a rather astounding 'third alternative' to contemplate with regard to the vital questions to be answered in the mystery of human origins and it is widely supported by recent TV programmes, highlighting the aforementioned popular culture and new age activists and their acceptance of it. Basically the theory holds that people should 'think outside the box' (or even off the planet) with regard to the quite amazing capabilities and achievements of the human utilising the over endowments of the human brain that set the human so far apart from any possibility of Simian Lineage. Such believers do not accept that these amazing mental endowments of the human brain were there by a divine creator as this would indicate (to use modern parlance) that he would have 'made a rod for his own back' by bestowing such amazing creativity that would only ultimately challenge the divine creator himself. This is borne out by the rapidly advancing strides in genetic science; the eventual creative capabilities of the human would eventually challenge the divine creator. The theory holds, that an 'extra terrestrial' force or influence was, at some point influencing greatly human development. This work, at the appropriate time, will highlight the 'evidence' albeit circumstantial that appears to support this proposition.

This work will dwell on, and make it clear by doing so, that it is a mistake to assume with regard to human origins that it is all resolved and that all the facts are in it will reflect on the purpose and presence of the human and its capabilities (on the positive side). In particular the human brain with its mysterious gifts and the very 'un-Darwinian over endowments of the human brain and its lightning development, capabilities and scientific achievements, that so obviously step outside the normal extremely slow processes evident in evolution and natural selection and what these mysterious gifts will allow us to achieve in the future.

As said, many people, through no fault of their own, falsely believe that all the facts are in and there are no mysteries connected with digging up old bones. Once this assumption is dispelled, the subject of human origins becomes a fascinating topic, entirely due to the mysteries that most certainly do surround it.

We are rushing headlong toward the ability to explore deep space, which of course, is a grist for the mill of the supporters of the aforementioned 'third belief' involving an extra terrestrial source to explain the human question, and that this was the ultimate purpose of ET involvement in the first place, to welcome humans to the galactic brotherhood. However, since it has become obvious that humans know where they are ultimately going, it would be nice to determine but not essential to resolve the mystery of where we came from.

For over a century and a half, with many energetic efforts by those active in the field qualified to assess the evidence, the question remains unresolved. The mainstream theory was first introduced around 1860 as a result of the joint efforts of Charles Darwin and Alfred Russell Wallace. This bold new concept as said was disturbing to many people, and shocked many into the consideration of the uncomfortable supposition that the hand of God may not, after all, be the reason for human appearance on Earth.

The situation that exists today is that many people largely pre-occupied and concerned with the pressures and priorities of everyday life seem to have neither the time nor the inclination to dwell on the mysteries of their beginnings. This is not so much due to their lack of interest but rather due to the fact that they do not even realise that it is a mystery and subscribe to the widespread belief that science has resolved it all.

Many people who must know better such as the aforementioned Richard Leaky surely must feel that science is lacking in its duty to dispel this assumption. Even though it takes the pressure off them to admit that they do not have all the answers and that all the original questions remain unanswered, in particular, the human brain and its capabilities that so defy natural selection and evolutionary concepts. An organ that so astounded Darwin's contemporary, Alfred Russell Wallace, as it defied all the usual evolutionary processes that all other earthly foliage and fauna appear to comply with, that it caused him to question the whole concept that he was happily in tune with. The end results was, he departed from the theory regarding the human connections and yet they could not quite put his finger on how to define this mysterious 'outside force' that could be offered today when we mentioned the third alternative (i.e. the ET concept). However, many believe, that all the facts are in and that all the vital bone fossil links that Darwin stated "Must be found or the theory falls down" have been found and placed on the fossil table. Furthermore, they have been subjected to references as "Our ancestors the primates" and "Our cousins the chimpanzees" for so long, that they cannot be blamed for assuming that science has resolved it all and that simian ancestry has been fully proven.

It could be said, that continuous references to 'ancestors and cousins' among simian kind, may be viewed simply as anthropological 'propaganda' but it is not really so surprising when we consider the fact that Charles Darwin himself stated that chimps and gorillas were man's closest living relatives, when his statements regarding human emergence where entirely theory and no convincing ancient human fossil remains had been found.

Naturally, the theory produced a great traumatic shock among the ecclesiastic fraternity and the shockwaves are still rumbling around today with continued debate and controversy, some quite heated in certain parts of the world. It has to be said, that science and discovery had, until more modern times, been viewed as 'satanic' in the dark ages and any brave scientific postulations sometimes resulted in death. It was a very disturbing revelation for the churchmen and the many religious people that fully subscribed to the teaching and events in the Bible, regarding their presence on Earth, to be told they were not created by the hand of God at all but ascended from apes.

Originally it was a nice thought to many people, to be told we were all created in some beautiful Garden of Eden by an all knowing, all loving God, but to be suddenly told that it is all untrue and our ancestors where apes, was decidedly uncomfortable. On the other hand, those who had no religion in the first place or were rather lukewarm in their beliefs, could tolerate it quite adequately and simply wait for the evidence to emerge from all the digging going on in Africa, as Darwin had suggested would probably produce the proof but as time went by, those most qualified in the research and study began to display their frustration, indicating in such statements as "Those who think that all the problems and issues regarding human origins are solved are simply deluded themselves" (F. Clark Howell), also "evolutionary studies (human origins) displays itself as more like a 'game' than science. There is much contention and dissent from within and without" (Prof. L. Washburn U.C.L.A.).

Nevertheless, the patient but rather frustrating search continues for the Elusive 'missing links' or alleged 'common ancestor' and that the elusive 'evidence paleoanthropologists all assume will eventually be found. The creature has been named in advance of its discovery as the 'pro-consul'. He is seen as the alleged common ancestor of both humans and modern apes. It is said that ape ancestry goes back to the earliest primates of the Oligocene or possibly the Eocene periods of forty to fifty million years ago.

As previously said many people asked and rightly so, why we can so easily find the bones, even complete skeletons, of creatures that roamed the Earth before the primates, then suffered the assumed elimination event of sixty-five million years ago that appears to have put paid to the dinosaurs? Our museums are full of such examples, yet the human remains to vindicate the Darwinian concept are conspicuously absent therein.

The problem of the missing fossil links has prevailed right up to modern times with no fossils to portray a distinct mutational change from the true pongid apes to the so-called 'hominids', apart from a couple of incomplete skeletons, the hominids are deduced from scraps and partially assembled skulls which all appear more pithecoid than human and may well be simply just another variety of primate, of which there are many. Bone fossils of the lower vertebrae and hip joints and skulls showing the lower point where the spinal column enters extremely rare and these are required to determine bipedal locomotion yet the paleoanthropologist seem convinced that the hominids where all bipedal. Certainly, the hominids appeared to have much more in common with gorillas than anything we could imagine as a pre-human ancestor such as the Aurignacion or Cro-Magnon peoples who were anatomically modern and appear to be our only convincing ancestor.

Anything preceding them, including the Neanderthals are questionable with regard to belonging to any convincing family free. The paleoanthropologist Don Johansen stated, "The variety of family trees now cluttering up the literature makes it impossible to identify the correct tree from the forest". He also stated that "Bipedalism did not guarantee humans".

Rather than millions, the true human ancestry could be measured in just thousands of years ago, an extremely interesting event took place at some point between fifty to one hundred thousand years ago regarding the human genome and will be related later in the work and assisted the lightning appearance of true humans that defied all laws of normal and excruciatingly slow evolution.
A being with a large and clearly over-endowed and developed artistic and creative brain. Charles Darwin was an enlightened and intelligent man who travelled (and sailed) widely in pursuit of knowledge with regard to the foliage and fauna of the Earth. Unfortunately, we cannot simply classify the human entity as just another earthly creature. Darwin seemed to have little use for a divine almighty creator that many believe waived an airy hand and everything appeared in seven days. He did however, believe in a 'force' that he never determined (although it had a name, natural selection and evolution of the species) a force that was ever scrutinising, evaluating, adding up and preserving all that was good and rejecting that which is bad.

That is of course a simplistic view and would be an evolutionary ideal. Unfortunately, this process is not reflected in the actual process. Science knows that mutations are extremely rare and when they do occur, they are predominately harmful, radiation for example and certain drugs used without testing such as the thalidomide scandal in the sixties when children were being born deformed or without limbs.

Of course, in Charles Darwin's days nothing was known of the double helix, the genetic code and chromosomes that determine any creatures characteristics and are relative to its species, and vitally protect that species from possible dangerous mutation due to interspecies copulation, in other words they appear to remain 'each unto their kind' . The fossil record shows that many creatures have remained identical to their ancient predecessors of millions of years ago (this is particularly evident in marine life) without the slightest change.

However, none of this applied to the human entity, which is certainly an 'enigma' and does not appear to have obeyed such rigid laws, particularly in regard to the genetic stability evident in all other creatures' insects and flora.

When it came to the subject of the human appearance, it is rather surprising that Charles Darwin could state in such a casual fashion that chimpanzees and gorillas were mans closet relative, when he must have known how it would disturb even infuriate the ecclesiastical fraternity, not to mention the masses of people who would have considered themselves as firm believers and accepted Genesis to explain human appearance on the Earth. He did however, state that he may have to think it out again; by making it known that his theory depended on the vital fossils to prove simian ancestry would be found.

We mentioned that Darwin's contemporary and one might say (in modern parlance), co-producer of the theory of Alfred Russell Wallace, waivered noticeably when trying to fit the human entity into the picture. He was aware Charles Darwin must have been, of the great intellect, mathematicians, physicists, astronomers and philosophers of their time and those who had preceded them and found it very difficult to simply assume apes were indirectly responsible for their gifted intellects. He felt that some 'unknown factor, or outside force or currently unknown variable must exist to explain the development of intellectual capacity in the human brain, particularly with regard to its amazing creativity. He corresponded with Darwin and made his concerns known. In response, Charles Darwin said, "I hope you have not murdered our child completely", (the theory).

When we consider theories, in general they conform to a routine of small incremental bits of knowledge and information collated to support the main theory than a thesis produced to be probed tried and tested in order to be able to stand up to scrutiny and the results must show repetitive factors, before it is put forward as a viable theory. One can see, in the case of Charles Darwin 'apes to men' theory, the process occurred the wrong way around, the theory was put forward with the assumption that the factors to prove it would be gathered to display gradual modification in human bone fossils, showing a transitional change from simian to human, after many skeletal structures had been found from different ancient time periods up to the present. Unfortunately, after much frantic effort the fossil table is disappointingly sparse with regard to ancient humans.

If Darwin was somehow able to come amongst us today he would be bitterly disappointed, but rather glad he had made his statement that stressed that importance of finding the convincing fossils to vindicate his theory. On the other hand, Alfred Russell Wallace would feel quite redeemed when seeing such major scientific advancements and plans to eventually visit the stars, all due to the greatly over-endowed brain and its capacity for mathematics, which would certainly qualify his reason for departing from the theory but in order to identify his puzzling 'unknown variable' he would have to analyse this other theory that has emerged and if doing so, he would quickly realise that this theory (not stated as a matter of fact in the hope that evidence would be found) but compiled from an abundance of evidence, written in ancient texts, and displayed visually, and this would most certainly qualify adequately as his 'unknown variable'. The plain fact is that human emergence and mental and intellectual development remains surrounded by mystery and unanswered questions.

In a book published by 'Gambit' in the USA, the author Norman Macbeth stated "As scientifically disciplined and qualified as those pursuing the search for the vital (human) fossil links are, they are slavishly adhering to the theory of the 'amateur'. Charles Darwin did not teach in a university or even work in a laboratory, he simply 'did' science in his own home without any trained staff and little fossil equipment".

I feel this is a little unfair to the man in the light of his worldly voyages and intensive research and documenting that contributed so much to our current knowledge. The plain fact is, "although we find vast deposits of the fossiled remains of now extinct animals, remains of 'ancient ancestors' are extremely rare and are of such rarity that the finding of a single fragment can cause worldwide headlines, sub-humans underwent a metamorphosis and became modern people. Perhaps as certain people have suggested visitors from another planet caused the change, or perhaps GENESIS is correct. Suddenly, intellectual and creative capacity occurred and the means by which this happened remains mysterious" (Ralph Franklin Walworth (U.C.L.A.)).

Chimpanzees, when standing upright for the short period their inadequate bone structure will allow, would not, one would imagine, be sufficient to have Darwin state that they and the gorillas where man's closest 'relatives'. However, he had been dwelling on the 'evolution verses creationism' since his young Cambridge undergraduate days, but it was still a large leap of faith considering the dominant following in his days of stern Victorian religious beliefs. However, there were many atheistic people in those times and those lukewarm in their beliefs and so, for them, when the Darwinian Theory exploded among the ecclesiastic, a favourable alternative was provided to account for human existence. It 'shook the tyranny of Rome', or more precisely the Vatican and what the late (reverend?) Ian Paisley rather irreverently called 'Popery'.

Blind faith was never quite acceptable to many lukewarm Christians of those times, but to accept Darwin's theory immediately without any positive fossil evidence was simply transferring one blind faith for another, but it has to be said that science has more of a chance to vindicate the Darwinian concept in the long run than the acceptance of the creation of everything being wrapped up in a week. However, with regard to explaining the enigma of the human, in particular the brain (in the best examples) science has had little success in the last 100 years to explain it in terms of the Darwinian concept of evolution ever scrutinising evaluating, rejecting, accepting and adding up all that is good etc. There just has not been sufficient time to produce the brains of such people as

Newton and Einstein. In order to study the mystery of human origins, one need not have been an anthropologist, just the capability to assess the evidence. One's interest in any topic requires one to evaluate as much data as possible, how else do we learn anything? However, whereas there are those in this field who display absolute faith in the Darwinian concept, there are others highly qualified in the field that are not afraid of being labeled a 'heretic' and tell how it is. Consider the following:- "Concrete evidence of the inadequacy of the Darwinian hypothesis is evident in the archaeological record" (Richard Leaky).

Nevertheless, in Darwin's time it was comparatively easy to change allegiance to his banner, in particular, those who were rather indifferent in their beliefs in the first place. Perhaps there would not have been such a noticeable exodus from biblical creation belief, if the Old Testament 'Genesis' account had not been written in such a manner, in particular the 'seven days' event but for this we have to lay the blame at the feet of poor old Moses who the British and Foreign Bible Society state was responsible for the biblical account. It was written by a fallible man not God. Equally, those who flocked to the Darwinian banner could only wait expectantly for the vital fossil links to be found, to show that men were once primates. Among the members of this new banner, were many men of science, who rejected any alternative theory to the Darwinian package stating "Theories must be tried, tested and evidence produced". Conveniently forgetting that they have embraced a theory that when propounded not a particle of evidence existed to support it.

Consider another statement by Richard Leaky "Holding up every old bone and pronouncing it as the vital fossil link exposes the folly of being chained slavishly to the entire Darwinian package all we possess is a meager fragmented array of single teeth and bones and portions of skulls. Studies of human blood groups shows no connection with the ancient apes. The important issue of human origins remains unresolved". The following is a quote from the British Museum Press of 1994 concerning an article headed "Human Remains". "Isolated fragments of bones and teeth can easily be miss-identified; fragmentary human bones are sometimes being confused with those of large carnivores such as bears ….. fragments of the limb bones of large birds such as swans or geese are also mistaken for human remains".

It is necessary to correctly date the rocks and strata surrounding organic bone material when such material seems to fall outside the 40 to 50 thousand range of carbon dating process. There are some interesting anomalies highlighted in my book *When the Moon Came* (Arena Books) when discussing the age of moon rocks brought to Earth by successive Apollo missions. They are taken from Don Wilson's book *Our Mysterious Spaceship Moon* (Sphere Publications). "After the N.A.S.A. analysis there were other report s, according to 'sky and telescope' a well know astronomical periodical, it was mentioned that the lunar conference of 1973 revealed that a moon rock had been discovered that was dated at 5.3 billion years old (older than our solar system). Another report, based on the potassium argon system of dating frequently used by science in determining the age of rocks claimed "some rocks given an unacceptable age of seven billion years" and stated "If we are to believe another report, two Apollo 12 rocks have been dated at twenty billion years old". All this shakes our confidence in the accuracy of the process. Today, the estimated age of the universe as eleven to fifteen billion years old. It can be seen therefore, that there is a great potential for error, particularly in regard to the many early bone fossils and skull portions found, and whether they were subjected (each portion) to accurate dating before being fashioned into complete skulls with the help of copious amounts of plaster and in particular the lack of advanced dating processes available. I refer to the fossils found by Richard Leaky's patients (Louis and Mary) in Africa.

We must remember that in those times, a strong compulsion existed to prove the Darwinian theory as soon as possible. The "God" of the Darwinian theory worshipped by the anthropologist and biologists is natural selection and minute changes over vast time periods and certainly, this scientifically orientated view seems to have more going for it than a belief that an all-powerful being waived a majestic hand and everything was accomplished in seven days. That concept could only be accepted by refutation of scientific logic and only maintained by blind faith.

However, as previously said, the hard to accept biblical account in Genesis was not written by God but a patriarch called Moses. Purposeful design could only (with regard to the wonders of the world and the universe) be (in the religious beliefs and concept) achieved by a designer. This 'designer' in scientific biological terms is, said to be 'natural selection'.

If a divine almighty creator did initiate life on Earth what was the point of the dinosaurs who lumbered about on the Earth for some hundred and eighty million years? Particular, Tyrannosaurus Rex the ferocious creature that ripped into the placid vegetarian eaters. Can we blame natural selection for them? If a divine creator did create them, why did he allow all of them to be wiped out and then wait for another fifty million years before the primates became modern apes then allow humans (in God's image) to be created roughly fifty thousand years ago, when anatomically modern humans appeared in a creation centre called 'The Garden of Eden'? This all contradicts the Genesis version with total creation completed in 7 days. We have to say that natural selection and divine creation are alternative belief systems one does not disprove the other. The eternal argument and dispute over these beliefs will seemingly rumble on until resolved one way or the other. The belief in a 'rapture' or when a divine creator appears in a 'second coming' where all the dead arise from their graves (or their jars) to be judged or damned still prevails, indicating how many people believe this. If it happens, it will be too late for all the atheists to rush to the church hoping for instant conversion. When we deal with the third theory for the human appearance on Earth, this second coming would be of quite a different nature.

With regard to the natural selection for the development of intricately formed plants such as the Venus flytrap for example, that gets its food by planned cunning, such a branch of life must have had a lengthy evolution with slow purposeful construction over very long periods of time, building and adding to the cellularly complete whole. This all fits in quite nicely with the acceptance of the Darwinian version "Evolution ever scrutinising, improving, adding up all that is good and rejecting all that is bad". It also fits in nicely with, and explains what natural selection is all about. All this is fine for plants but fossil discoveries indicate just what the Bible says with regard to some foliage and certainly the animals, they have remained each unto their kind, this also includes, flies, insects, frogs and in particular marine life with very little change over millions of years. A good example of this is the 'Coelacanth', a marine life that was found in a fishing net and was thought to have died out many millions of years ago. A first class example of genetic stability evident in all living things except the human entity. Alfred Russell Wallace, Darwin's contemporary, who wrestled with the enigma of the human, in comparison to evolution in all other things, sought to find some other unknown factor that could explain the rapid evolution and over-endowments clearly displayed in modern human anatomy and brain development, especially in comparison to the poorly endowed apes. He knew evolutionary processes never over-endowed any species in advance of their needs to survive. In the human brain, only a third of its cellular material produced the great philosophers and mathematicians, it is staggering to imagine what it will be capable of when (or if) the rest of its structure becomes active.

The 'force field' from the human brain's explosive development, over such a comparatively short time, blasts both natural selection and plodding evolution out of the water. The human brain in defiance of evolutionary processes blossomed like a mushroom between night and morning in evolutionary timescales. We mentioned a third contender for acceptance along with Darwinian evolution /natural selection and divine creation, which would certainly adequately explain this mystery of the over-endowments evident in the human entity and will be fully dealt with in this work. However, with regard to the brain's not so astounding qualities, such as criminal and murderous activity and the obvious negative features, within the brain, all disproves any notion of infallibility of any alleged extraterrestrial beings or a divine creator, and we certainly cannot blame any primate 'ancestor' for them. When we mentioned orderly and slow cellular construction of flowers and foliage in the evolutionary process, can natural selection explain the massive explosion of marine creatures during the Cambrian period? Some six hundred million years ago, creatures appear with no fossil history, this even perplexed Darwin himself and he admitted that he had no answer for it. Although there is a third contender that has been introduced in this work, the two main theories for the appearance of the enigmatic human and its over-endowments in terms of intelligence are the Divine creationists and the 'natural selectionists' that is, those who support the Charles Darwin/A.R. Wallace thesis. The creationists have it that a divine God was responsible who is often referred to as "Our Lord" so for this analysis, we will call him Mr Lord but who is Mr Natural? We know what selection means humans can 'select' but we must try to identify Mr Natural. Mr Lord declares, "I am the one who is responsible for the beauty, wonderment and obvious design that you see in my handiwork". However, Mr Natural cannot deny the existence of Mr Lord anymore than Mr Lord can deny the (shall we say handiwork rather than existence) of Mr Natural. An atheist cannot prove a divine God does not exist and cite natural selection as any kind of proof. Mr Natural and Mr Lord are both contenders for the same accolade, both claiming their right to it. We can only decide who we think is more deserving, rather like choosing one of two candidates in a bi-election who speaks with the most convincing theories but we cannot deny that profound beauty and design does exist. The amazing varying designs evident in a snowflake or the intelligence and one could say 'cunning' in some plants such as the aforementioned Venus flytrap that ingests insects or those that invite them in, such as bees so their pollen can be transported away to other plants for their propagation. The telegraph vine is said to be able to signal to others of its kind many miles away. Although there is obvious intelligence in plants, some people advocate that if you talk to them in a kindly way they flourish more than those you ignore. So, if for example, you get up close and personal with your tomato plants, you must realise that you are providing them with more carbon dioxide than those you ignore, therefore they will flourish and that it is nothing to do with plants having emotions and a preference for flattery.

In terms of preference many people reflecting on the contenders, that is Mr Natural and Mr Lord, would surely prefer a mighty and magnificent 'divine creator' rather than cold impartial Mr Natural, but in the final analysis it still comes down to a matter of preference or choice.

For all that, we mentioned a third contender; he is a 'cosmic tourist' and sets himself apart from Mr Natural and Mr Lord. He does not claim to be the designer of all the beauty and wonderment of nature but he makes a very strong claim for the wonderment evident in the human brain and its capabilities and over-endowments that the humans will eventually utilise in order to become cosmic tourists themselves. On this point, Mr Natural and his team never over-endow any creature in advance of its needs to survive. In the Einstein's of the World, the human brain is massively over-endowed yet (shall we say) ordinary folk also possess two-thirds additional cellular structures supplied (by whom?) but currently unused. It must eventually have a purpose, which no doubt also keeps Mr Natural and his team guessing with the rest of us. Mr Natural must find it very difficult to explain how positive mutations from a primate, swarmed with such force and numbers to provide the human with all that intellect and the important capacity for mathematics that the humans will eventually utilise fully in order to become cosmic travelers themselves in any 'natural selective' way. We have reached the point where Alfred Russell Wallace waivered and pondered on his 'unknown variable'. Here, the aforementioned third contender or the possible cosmic tourists may well provide the answer if consulted.

Loosely, evolutionists see the line of human development as pro-consul, hominid, Homo erectus, Neanderthal and Cro-Magnon (our only convincing ancestors). If all the preceding entities where supposed to be a successful mutation toward the human why did they all expire? They are all questionable as belonging to any human family tree.

Unless a recent discovery has revealed otherwise no links exist between the true pongid apes and former pro-consul said to be the forerunner of the hominids and the so-called 'modern apes'. He is an imaginary figure at this point in time and a wide disagreement exists as both the time of his alleged appearance. As for Homo Erectus, he was said to have walked upright but had the skull shape (but more brain material) than an ape. All, most probably different varieties of primate, of which there are a great multitude. Upright primates are perfectly feasible. Ostriches, ducks, geese, hens and birds in general all get around on two legs even though Homo Erectus was said to have walked on two legs the paleoanthropologist Alan Walker stated (with regard to Homo Erectus) "There was no human consciousness in Homo Erectus, he was not one of us and did not think like a human and had no language capability. The Cambridge University Archaeologist Paul Mellors stated "Nothing resembling Neanderthal D.N.A. has ever shown up in humans. All this seems to refute any notion they where our ancestors and raise more questions that require answers, in this strange disjointed array of alleged human predecessors. As said, Charles Darwin assumed that Africa would be the place to produce the vital fossil links that he needed to be found to finally (in modern parlance) put his theory 'to bed'. One can only imagine he thought this after stating chimpanzees and gorillas where mans closet living relatives as they existed in such numbers there. Much hard and uncomfortable work then began in that land. There was a lot of activity regarding the search during the 50s. Richard Leaky's parents Louis and Mary where very active in this regard and seemed to have more ardent conviction of the Darwinian concept than their more conservative son Richard. The account in *The story of archaeology* (Wiedenfelt and Nicholson 1996) mentions an event in 1959 where Mary Leaky was walking her dog and some skull portions where 'accidentally' discovered. They were then re-buried until a film crew arrived to record the event. Interestingly, it was stated, that a discussion between her and her friends revealed a previous intention to make the film. Perhaps because the date signified the publication of Darwin's apes to men theory propounded in his 'on the origins of the species' exactly one hundred years previously. How convenient was that? These fossil finds where discovered at the site of Olduvai Gorge in South Africa. Eventually a skull was constructed from 400 pieces of bone, found with an abundance of animal bones. Now, one is not suggesting that the good lady filed a bit here and ground out a bit there to assemble it, but it did require generous amounts of plaster to put all the pieces together but we have to ask was every piece subjected to testing by the processes available at that time to ensure that they were all of the same age and came from the same creature the potassium-argon system of dating had come into force at that time, which was capable of dating organic fossils further back in time than the limitations of the carbon dating system? This new system of dating revealed the scandalous hoax of the Piltdown 'man' which will be related in detail later. The end result was that the skull that was cobbled together from so many pieces, more resembled a gorilla than anything human looking in appearance, it had a large sagittal crest and lowering brow ridges and only the teeth seemed to link it with a supposed human predecessor.

The long list of paleontologists, biologists and archaeologists that have come and gone have unfortunately and most frustratingly been unable to unearth the vital fossil links, yet those (to use a Richard Leaky phrase) that are "slavishly chained to the entire Darwinian package" seem to have ensured that the theory changed gradually from a theory into a widely accepted 'dogma' and made it appear quite unassailable yet one cannot say it rests on solid foundations.

Many 'men of letters' and impressive qualifications have made errors of judgment, mistakes and highly contentious claims and statements in the field of paleoanthropologists and their other sciences employed in the search for ancient ancestors. However, impressive qualifications are not a pre-requisite for one to have an objective study and research the books journals, facts and data currently available in any good library that have been patiently compiled by those directly involved in the field, concerning the search for the elusive ancestral links. One simply needs a reasonable education, good reading ability, an open mind and most of all an avid interest in the subject. Almost all our knowledge is obtained through reading and since the necessary books and journals are available to the public regarding this topic, one only has to go and research it for oneself.

Most anthropologists writing in such works will admit that the theory we are dealing with has not been proven beyond doubt to everyone's satisfaction and all the facts are most certainly not all in. As it, stands there are said to be no fossil links between the true pongid apes and the disputable hominids and the 'pro-consul' would fill this gap quite nicely.

Many unanswered questions and many problems, anomalies and widely diverging opinions exist throughout this contentious field of study and doubtless there are those that in spite of their unconcern at the prospect of having Simian ancestry, do object somewhat to the projected air of unassailability that surround the theory and the constant references to 'cousins' and 'ancestors' until the point is reached when the theory is proven beyond doubt.

Most anthropologists dislike the term missing links, primarily because such terminology only serves to remind them (understandably) of their irritation and frustration in not being able to discover the vital bone fossils when such mammoth effort has been applied to the search, quite often in hostile conditions (weatherise) and great discomfort to themselves.

Currently there is not a single skeleton of a convincing ancient human ancestors that could be described as one hundred percent complete when one would expect our museums to be full of them. There is an entity that was discovered near Lake Turkana in South Africa. The creature did not appear aged and was called the 'Turkana Boy'. It was assembled after sifting through fifteen hundred tons of sediment over various seasons in the field. It had an ape-like skull and was classified as a Homo Erectus entity. Another creature discovered in the 1970's that was named 'Lucy', as the Beatles hit song, 'Lucy in the Sky with Diamonds' was popular at the time. It was alleged to be female (hence the name) but the usual dissent prevailed and this was hotly disputed. It was classified as 'Australopithecus Afarensis' and was claimed to be an upright walker but because it had curved finger and toe bones that seemed to suggest a tree living existence again the dispute prevailed, it was around forty percent complete, even its gender is not agreed upon.

It is a strange paradox with all the disagreement of opinion in the field of paleoanthropology and the distinct lack of bone fossil evidence and the interpretation of all the bone fossil finds, that there is universal agreement in the belief of the apes to men theory and that it is unassailably correct. The fact is that the entire theory has largely been in a state of flux since its introduction, simply because there is no good reason why we cannot find correctly interpret and then display for all to see a whole series of complete skeleton frames of ancient and quite recognisable ancestors that appear to be slowly gaining modern anatomical features. To recall Charles Darwin's 'escape clause', "If the fossil links are not found the theory falls down". As, clearly, they have not been found after such purposeful and extensive searching, many people would feel justified in asking is the theory falling down.

We must remember that in the stern Victorian era, many people, not to mention the church where outraged at this attempt to reduce the 'noble human' to animal lineage. However, if ones leaning where more atheistic or indifferent they could bear the brunt of this theory and embrace it without any qualms.

Over the lengthy period of searching, if a bone was found, in a particular skull portion quick calculations where made as to the capacity or volume of the skull and brain size, if it was significant it could generate worldwide headlines and the graphic artists where commissioned, the end result was an obvious upright human entity with only vague simian features, spear in hand and nobly hunting its prey. It would have been more advantageous if the lower portion that of the jawbone, assembly had been found along with the few ancient skulls we possess, rather than just the upper portions. Then, we could see the point, where the spinal column enters the base of the skull (foramen magnum) and the further forward this is, indicates upright stance but of course, we hear anthropologists stating, "Bipedalism did not guarantee humans", although others disagree. Some anthropologists have disputed whether the various Hominid creatures could have interbred. By far the most important mystery is the rapid evolution of the human species, in particular the last one hundred thousand years and the complete and utter disregard that the emerging human entity seems to have had for the otherwise rigid laws of genetic stability evident in other creatures who remain 'each unto their kind' for millions of years. This also applies to foliage remaining the same as their ancient counterparts as the fossil record shows.

Those proficient and active in the field are aware (particularly the biologists) that the chromosome pattern in all species is different and specific and guards against the dangers of contamination and mixing, resulting in chaos among the species. This also applies to our alleged 'cousins' or 'relatives'. We could never produce the kind of creature they are all searching for simply by infusing a chimpanzee with human reproductive material and expect any results. Nevertheless, in the future, with the rapid advances in genetic science, this kind of manipulation may well be possible, and it may well have been possible in the past (but not by human hands) and may have resulted in our very existence when a certain group 'not of the Earth' stated "Let us make men in our image" which is stated in the very book Darwinian supporters totally reject.

To illustrate just how incredibly slow and genetically stable life forms in general are is to contemplate on the fact that six hundred million years ago most of the primary classification of marine life that we know exist today, were apparent then and show up in the fossil record. These are sponges, sea lilies, star fish worms, water fleas, shrimps, clams and so-forth, and after this enormous period of time including fish in general they look pretty much the same. Yet here we have the human in this intriguing period of around a mere fifty thousand years ago, going through enormous genetic modification and changes, in tissue building, muscle and sinew re-routing, a huge increase in brain material (most of which is currently unused) and bone re-structuring, almost as though the 'unknown variable' that A.R. Wallace sought, was actively genetically engineering the human body form. All this totally defies natural selection and standard evolution that required all those millions of years with little noticeable change.

There is no really satisfying explanation for these strange factors. In order to work therefore, it seems that the entire theory of evolution and natural selection has to defy its own laws. Although genetic stability rules and each individual species has its own distinct chromosome pattern, evolution is happy to condone a wide variety of types within a species such as the dog, or the primates, said to number some 92 different varieties and the possibility exists that the hominids, homo erectus, Neanderthal where simply other varieties of primate, whether bipedal or not, yet quite unsuccessful. The anthropologist Don Johnson (the discoverer of Lucy) said "Bipedalism did not guarantee humans so it is perfectly possible that bi-pedal apes with other differences in teeth and so forth, could have emerged and hung on without any improvement precariously for the comparatively short time that the fossil evidence suggests before meeting their demise. In all this, the true pongid apes continued their long evolution, said to be some fifty million years since the earliest primates, yet show no inclination toward improvement or change toward any humanlike features and seem quite happy to remain as apes. Whatever advantages hominids, homo erectus or the Neanderthals may have had, they provided no help in preventing their complete demise while the pongid apes still flourish.

The facts that there is only one distinct and anatomically modern skeletal frame in existence for Homo sapiens that remains the same for its entire existence, once thought to be only thirty-five thousand years ago but now pushed back to one hundred thousand years. However, the monumental event of some fifty thousand years certainly requires a very careful and convincing explanation, which will be mentioned herein. All the preceding entities are the subject of contention, mystery, conjecture, questions and anomalies and are represented by mostly scraps of bone material.

It is all these factors that make the subject of human appearance an interesting and fascinating topic and of course makes the theory of the so-called 'third alternative' for the appearance of the modern human equally intriguing, which is interwoven within the pages of this book and goes somewhere toward clearing up many questions. Clearly, the paleontologists must be in the most frustrating profession on Earth in being unable for so long, to discover the vital evidence to vindicate Darwin's apes to man theory once and for all but of course, this could happen any day, assuming the search is still as intense as it was in the days of Louis and Mary Leaky and their sons Richard and apparently with some effort from another son Jonathon.

There is also a huge time span in which the anthropologist are divided, and that is the initial appearance of the alleged and long searched for Pro-Consul, and it ranges from five to fifty millions years, depending on their individual viewpoints. In any event, his actual appearance would have had to have been quite a monumental occurrence simply because he would have had to defy the afore-mentioned chromosome pattern of his original species and mutate or change to a different species with a new set of chromosomes. Moreover, it is well known that mutations are, for the most part, harmful and disadvantageous in a species and so the most powerful circumstantial evidence against the possibility of his appearance would have been his unlikely ability to have pulled off a 'double whammy' so to speak, in not only defying the chromosomes pattern of his former species, but to have achieved a successful mutation into the bargain, and of course, the studies of human blood group not being able to prove any obvious Simian connection.

The simple fact is , that all the previous entities, *before* Homo Sapiens, if they were a mutation, were all very unsuccessful events, simply because they all died out and disappeared completely off the face of the Earth and, if they were a positive link in the lineage down to modern Homo Sapiens, then their gene pattern would have some slight effect and make itself apparent in the future species of their descendants from time to time and studies of mitochondrial DNA, inherited only from the female or mother, make it clear that this did not happen.

Although most people today dwell little upon it, when the bold new theory of evolution from the primates was first introduced, it must have been quite a disturbing and monumental revelation in many ways, particularly to very devout people accustomed to their former and well entrenched beliefs in divine doctrines. But then science has been disturbing the church for centuries, ever since the earlier astronomers dared to suggest that the Earth was not the centre of creation.

Nevertheless, when the Piltdown fraud was discovered by more modern scientific processes and it was realised that human skull and ape jaw portions had been purposely buried to be found and classified as the famous 'missing link' many people must have been extremely distressed when realising that the fraud remained undetected for forty years, from 1912 until the fifties, simply because more than a few people may have, perhaps reluctantly abandoned their former beliefs in favour of this new and convincing 'evidence' and, perhaps, lived and died holding this false assumption. They would have changed their allegiance to a theory that still remains unproven today.

Furthermore, this outrageous fraud even caused consternation and distress among the paleoanthropologists who referred to these fraudulent finds as a type of yardstick, or reference point in their subsequent finds and classifications. Raymond Dart's discovery in 1925 of the so-called 'Taung Baby', a skull which other anthropologists stated was simply 'the distorted skull of a young chimpanzee', was found when this Piltdown fraud was still accepted as a genuine fossil find. It was discovered at Taung, hence the name, at a quarry site in South Africa. The point in question was that this discovery was not only disputed in its interpretation and significance (like most other discoveries), but was rejected on the basis of comparison with the Piltdown fraud that was still ultimately to be revealed over twenty-five years later, as a hoax. And so, this monumental deception not only confused and disturbed those with strong religious views, some of whom would have gone to their grave believing they had made the right choice in rejecting their former beliefs, but it had an equally disturbing effect among the paleoanthropologists and biologists as well.

Many people today with no particular axe to grind and not inordinately religious, with an open mind on any theory for human origins on their merits and evidence, proven or circumstantial, still remain unconvinced by the so far discovered evidence for an otherwise acceptable theory.

As we have mentioned, Darwin did not base his theory on any puzzling fossils discovered in his time, so we must task, did he fall into the trap of assuming Simian ancestry because of the vague similarity chimpanzees displayed when standing almost upright and seen from a distance? Their incorrect hip joints will not allow this for long, as they are more suitable for knuckle walking on all fours. Chimpanzee's hands, when looking at the thumb, are not really like a human, nor are the rest of their body form for that matter.

It may be interesting to know that medical scientists consider pigs to be more akin to humans with regard to bodily tissue, and naturally, they are aware that pigs cellular material has been utilised in heart operations, but how many people are aware that British Army Surgeons in Denmark (NATO operational training?) had practiced field surgery techniques on anaesthetised pigs, that had been purposely shot then operated on to remove the bullets? Perhaps it was carried out in Denmark, as ethical restraints may be less severe there. However, one would have thought that an alleged 'cousin' would make a more suitable donor but the reason why pigs were used was (it was stated) because 'their physiology is so akin to humans'.

At one point, six or seven hominid varieties existed to be studied regarding the skulls, with a noticeable absence of the rest of their bones and their dentition seemed to point in a different direction and their brain cavities differed but hominid dentition was a great deal different from humans. In the Primate array there is a markedly different size in the brains of spider monkeys compared to gorillas and many people with regard to the sparse collection of hominid bones could see nothing to prevent them being classified as just another variety of primate. In any case, nothing helped to prevent their demise, and it seems clear that their chances to avoid this would have been better, if they had remained as apes rather than having pulled off this mysterious but detrimental mutation to become a hominid.

As previously said, anthropologists have stated, that the anatomically modern Cro-Magnon type were contemporaneous with the brutish Neanderthal and with regard to their attributes, it will be shown just how modern they were for their time, in the appropriate chapter, also, it is hard to imagine a Cro-Magnon male being attracted to a female Neanderthal as they were already degenerating and becoming more squat and beetle browed that their earlier counterparts, and their bone fossil samples clearly portrayed this, which is hardly an evolutionary trait, where the 'ever improving and selecting' process is supposed to prevail, so it does not seem likely that the two species interbreed or would have wished to.

It is said that no evidence has been found in any of the excavations of 'warfare' or violent aggression between them. Neanderthal bones have been found in strata below those of the more modern entities. Cro-Magnon came and the Neanderthal bowed out as though clearing the way for their more advanced successors. With all this, it is not really so surprising that an anthropologist stated, "Human evolution seems more like an 'experiment' than a naturally improving process.

All the various primates have happily existed in their current bodyforms for millions of years and we struggle to find and identify any creature along with the primates showing a distinct change toward becoming something else in the bone fossil evidence. "Each unto their kind" prevails and such changes would obviously have to occur in order to conform to the cartoon depiction frequently shown, to convince us of the evolutionary progression. I refer to the frequently depicted process of a fish dodging predators and struggling out of the primordial sea, turning its fins into legs, becoming crocodile shaped then a dinosaur, a primate who is shown slowly rearing up on two legs to finally shape shifting into modern humans.

There have been some very odd skull forms found that had all the necessary human features except for two vital differences, they were elongated and most importantly, had no 'sagittal suture', that is the split that is evident in the skulls of babies, that gradually closes up with age and maturity. These skulls have yet to be thoroughly studied. They will prove to be quite a challenge for anthropology. They have been found in Peru, South America, Malta, Egypt (shown in Egyptian Glyphs) and in Northern Europe, even in Germany and Russian, they are noticeable longer upward and to the rear. Of course, major differences in human types exist in other features of the human body form and skin colour facial characteristics and so-forth. Consider the seven-foot Watusies pygmies, natives of the rainforest, Chinese, Negroes, Japanese, Polynesians and Caucasians yet all are human and all can interbreed.

For all that, the mysteries surrounding the appearance of modern humans still prevail. When we mentioned the Neanderthal replacement 'by Cro-Magnon, in the 'Book of Life' (E Bury Hutchinson) Stephen Jay Gould stated "The Cro-Magnon peoples where contemporaries of Neanderthal for a time but not their descendants."

There have been great eliminations and wipeouts in the past, evident in the fossil record, and it is stated that a rather staggering 90 plus percent of all creatures that ever lived have become extinct. If this happens on a world so evidently conductive of life in general, when we only have to look around us and realise how quickly we could be swamped by prolific growth of grass, weeds and foliage. If we neglect to deal with it, it is not surprising that the chances of life having a firm foothold elsewhere in space could be rather remote. Most of the planets continually being discovered appear to be unsuitable for human habitation.

Our part of the galaxy could be viewed as rather a safe haven compared to some areas in space, where our astronomers are witnessing events that would seem quite terrifying to beings occupying any nearby worlds, with stellar explosions, galactic collisions, etc., yet around two hundred and fifty million years ago on Earth, the fossil record tells us that almost every living thing on Earth was catastrophically wiped out, and so our frantic technological development would seem not only to be exciting and interesting but downright essential if we are ever to save ourselves from any future catastrophe, as there is no reason to suspect that they are all in the past. It would seem that humans have been fortunate to have avoided the major events.

Those who perpetrated the Piltdown hoax, although possibly not considering the profound effects of their deception, were probably motivated in no small way by their dedication and belief in the Darwin/Wallace proposition and were subconsciously portraying their impatience and frustration with the obvious lack of realistic and convincing evidence that was failing to be discovered which, at the time of perpetration, had reached a point of some fifty to sixty years after the theories of Apes to Men first began to be considered. By perpetrating the hoax they were displaying the same motivation evident in the master forgers of the Middle Ages, who could produce such convincing artifacts that they may *still* be astounding us today, if the disputed Turin Shroud, an issue that has not yet been settled, does turn out to be a hoax, or some of the holy relics now possessed, or pieces of the genuine Cross and so forth. The forgers of the middle ages where extremely skilful in their work.

No doubt modern day fraudsters could, if they so wished perhaps produce very convincing fossils that had been subjected to certain processes, or artificially aged by some scientific method on a par with the same scientific advances that discovered the Piltdown fraud. It is possible that this has already occurred but they have not been unearthed.

While on this topic, an interesting question might be, when the new processes, such as the potassium argon method of dating which came into being in the fifties and revealed the Piltdown hoax, how many *other* finds, discoveries, claims and skull portions was it applied to at the time? There must have been a great fear evident among the paleoanthropologists of the time that other evidence of fraud could come to light, or even incorrect age interpretation of other finds. One could envisage a certain understandable desperation in the early days, after the introduction of the theory, for it to be proven as quickly as possible. After all, decisions to abandon former beliefs would not have been taken lightly, except by those who had little belief in the first place, and at the time, there were very few other alternatives to turn to. No proposals of 'biological accidents', or 'alien biogenetic creation' existed in those times and, consequently, those changing allegiance to this revolutionary, if not disturbing, new theory, would have nowhere else to turn and no 'get out' clause or 'deals' done with the church, that if the fossil links are not found then they would return to the church. Of course, many people are saddled with this dilemma in the present era.

Consequently, there was no going back and they were stuck with it and are still in limbo today, with reference to the still elusive missing links that continue to plague and frustrate them in their ability to discover the Pro-Consul. The current fossil finds are subjected to the carbon 14 method for organic substances and/or the potassium argon process for dating material and strata adjacent to the find, or the T.L.P. (thermo luminescence) process but has been stated that some scientists doubt the 100% foolproof reliability of *any* of our current dating methods.

This compulsion or desperation, evident in the early days to prove the theory, must still exist to some degree today among the so-called 'bones and stones' men. This could be apparent in the ego driven urge to achieve some sort of notoriety and fame, in first the newspaper headlines, then in subsequent books and journals, and a yearning or longing to be the discoverer of the famous missing links and to be remembered in the history books for their deeds. It is this kind of drive that causes such disagreement among those struggling to achieve this honour and depicts them more as rivals than contemporaries.

An early edition of *Time life international* detailed the kind of claims that were being made with regard to the current human fossil finds in the twenties and thirties. It also mentioned the fossil discoveries found in the Neander Valley in Germany (now classified as Neanderthalensis Sapiens). People intent of exploiting this lack of evidence such as quarrymen and farmers saw the chance of monetary gain by unearthing old bones.

History, both past and modern, has proved that scholars, for all their impressive academic qualifications, can be wrong. Rudolph Virchow, who was the leading European pathologist at the time of the early Neanderthal bone fossil discoveries, assumed they were simply from an ugly example of a human and dismissed an upper skull portion that was found in the Neander Valley, which was quite thick in bone structure, with massive 'lowering' ridges over the eyes, as "A not very ancient pathologist idiot". Another German authority of those times theorised that the remains could be those of a "Cossack that came over from Russia with others (probably to take part in the battle of Napoleon's retreat).

Even Thomas Huxley, who enthusiastically beat the drum for the Darwinian concept said of the finds, "Under whatever aspect we review[1] this cranium, whether we regard its vertical depression, the enormous thickness of its supraciliary ridges, its sloping occiput, or its long and straight squamosal suture, we meet with apelike characteristics, stamping it as the most pithecoid (apelike) of 'human' crania yet discovered, but he still referred to it as 'human' but his description does not warrant such a conclusion.

In the article, Huxley concluded that the Neanderthal bone fossil skull was more nearly allied to the higher apes than the higher apes themselves are allied to the lower. In other words, what Huxley was saying was that Neanderthal's skull was more akin to the apes than some apes are to others within their kind. This, of course, highlights the point that there are many differences in the wide variety of primate varieties, including spider monkeys, chimps, baboons, orang-utans and gorillas, not to mention the host of other primates, such as the lemurs, tarsiers and bush babies and so forth.

For all that, certain articles have claimed Neanderthal was 'a man' and a most successful one, as he dominated Europe for thirty-five thousand years, before Homo Sapiens took over. Many would question the view of Neanderthal being viewed as 'successful' in Europe when only lasting for thirty five thousand years, which in evolutionary terms, especially when considering a mere frog looked the same one hundred and seventy million years ago, is but a blink of an eye.

Man is supposed to be the prime species with dominion over all others, so if Neanderthal was classified as a 'man' he was an abject failure. The problem is highlighted and compounded even further when we consider that the Hominids, a supposedly less advanced and more pithecoid entity, lasted for many more years, with smaller brains and distinctly less humanlike attributes than the Neanderthals.

Paleoanthropologists, although rarely agreeing with others and their opinion and interpretation of their fossil finds, all unite in this contentious field regarding human evolvement from the primates and continue their long patient and dedicated work in the assumption that one day some profound discovery will settle the issue once and for all. However, unless such a find proves to be indisputable there is every likelihood that the same dissent, disagreement and counter suggestions will prevail.

Until this significant discovery does occur, assuming of course, that the required bone fossils *are there to be discovered,* it is questionable whether paleoanthropologists should continue to use such phrases as 'our ancestors, the primates' and 'our cousins, the chimps'. It would not be an exaggeration to say that those people who accept in popular culture the possibility of ET human creation, it would be the same as continually stating 'our creators, the extra terrestrials'.

The cover leaf of the book from *Lucy to Language* [2] says "Part one addressed the central issues facing anyone seeking to decipher the *mystery* of human origins". Here is a fine glossy hardback work by a renowned researcher in the field, the man who discovered the partially complete skeleton named Australopithecus Afarensis, christened 'Lucy', certainly, from the very onset, makes it clear that human origins *are* a 'mystery' to be solved, and of course it is quite clear that it is such.

Certainly, many factors qualify it for that description. In the earlier days of the search, when it seemed so urgent to prove it all, many paleoanthropologists, eager for the publicity and notoriety that would not do their careers in this bold new science any harm at all, all longed to be the other one who would discover the vital and compelling fossil links and, as a result of this, a new human genus seemed to be declared every now and then, with almost monotonous regularity. In addition, if they could get it to survive, the varied opinions of their contemporaries (one could almost say rivals) and have it established into the family tree, then it was duly logged, named and accepted.

The actual find promoting the new genus may only be *a single tooth*. However, once the discoverer has his find accepted and established, after a very short time, the anthropologists begin to make breathtaking assumptions as to the appearance of the entity that was alleged to have possessed it, and commission the graphic artists to depict a quite human looking entity up on two legs and nobly hunting his prey.

This certainly happened after a few skull portions were found in England in an area called Swanscombe in Kent in 1935. These portions became the Swanscombe 'man' and were dated as two hundred and fifty thousand years old, predating Neanderthalensis Sapiens by over a hundred thousand years, yet depicted as a modern looking human. Could this be viewed as an example of the 'scrutinising' selecting and rejecting and 'ever improving' that we hear of in evolution and natural selection parlance? A more advanced anatomically modern entity superseded by a squat, grunting sub-human type, where evidence of cannibalism is found among the Neanderthal bones? How can the evolutionary processes jump about like this? It is little wonder that the human fossil record conforms more to an experimental process. If, as previously said, it was possible to find complete ancient skulls with the base intact and the lower vertebrae and pelvis assembled and accurately date them, a better picture would emerge with regard to the Hominid types, but without the Pro-Consul discovery it would still remain impossible to ascertain the precise date when this monumental change, or mutation, took place, to begin the long strides from true Pongid apes to humans that remains completely unresolved today.

If we look at this event closely, we have already established how profound it would be to achieve this 'double hit' in changing species and successfully overriding the chromosome pattern and achieving a successful mutation. However, the Pro-Consul has much more difficult tasks ahead of it. It has to produce in its offspring modern apes on the one hand and humans on the other, since it is dubbed as the common ancestor of both apes and men. In other words, its offspring would have to take two distinctly divergent paths. Firstly, surely there would not be a continued requirement to produce *more* apes, when, if postulations of Pro-Consul's appearance are in the *lower* scale of assumption, that is around five million years ago, simply because the planet was already replete with apes and primates and of course many primate varieties has existed since the Oligocene, possibly the Eocene periods.

If we consider the two divergent offspring recently born of the earliest Pro-Consul (we must surely have to assume male and female Pro-Consul), we would have to envisage two distinct behaviour patterns in the offspring. On the apelike path the infant could very shortly be jumping upon its mother's back to ride around on her while she foraged for food. However, what about the infant on the human side?

I said in another work , that if the human suddenly adopted the practice of the animals and left its 'cubs' on the ground after birth, the human race would cease to exist. This helplessness of the prominent species on Earth has never been satisfactorily addressed. It seems to suggest that humans are the product of an enormously long evolution (that does not seem to have been possible on Earth), where another guiding force, or 'machine technology' has taken over the administering of the needs of the newborn infants, making them helpless and dependent on it, or appear to be 'born too soon', which would not appear to be a desirable trait. So the question obviously is, how did we survive to become the dominant species in the first place, when we would, with such disadvantages, have been earmarked for extinction long ago? Four legged creatures immediately struggle and quickly rise up and find the life-sustaining teat with no help from the mother but the human infant has to be lifted and guided to it or it would expire.

Some anthropologists adopt the strange logic that *all* humans are hominids, yet at the same time agreeing that not all Hominids could be called humans. Many people would go further and say *no* Hominids could be called human. If it was possible to produce one in some kind of Jurassic Park cloning experiment from DNA, there would be no better place than in the trees or one of the zoos for them.

All the Hominids are classified by their dentition and their huge grinding molars as 'old world monkeys and apes' , and have sixteen upper and lower teeth, with the aforesaid large powerful molars. Humans, on the other hand, are distinctly different with regard to their dentition, with fewer and less powerful teeth and more importantly (and distinctively), a complete lack of 'diastemata' or gaps between the teeth. This gapless feature of the human dentition is unique only to humans, one of many unique features in humans in comparison to animals. As opposed to a 'wooden' stare, humans can practically speak with facial movement. An eyebrow cocked, a mouth turned down, eyes widening or narrowing etc. Humans, with the profound gift of speech, have the larynx low in the throat in order to achieve, it and they achieve it with the help of an enormously over-endowed and developed brain that also completely defies the laws of evolution and natural selection. This brain has a specific location called the Brocas area, situated in the fore brain, the area that has astounded anthropologists ever since it equally astounded Charles Darwin and Alfred Russell Wallace. Amazing departures from normal genetic stability were necessary with the advent of our immediate ancestors the Cro-Magnon and quite profound alterations of the skull shape in order to accommodate the fore brain to provide our distinctive human qualities. All the Hominids had the thorax in the chimpanzee position, or high in the throat, as indeed do all mammals.

From all that has been said to classify the human entity as a Hominid, would seem very questionable indeed. According to the late Sir Arthur Keith, the medical anthropologist, the human has 312 distinctive 'generic' traits unique to its species, as opposed to the gorilla with 75, and the chimpanzee with 109, and he said, "Whatever theory is propounded for the origin of the several members of the higher primates, it must account for their structural and functional characters". [3]

In terms of the most unique traits in the human entity, they are breathtaking in comparison to any other animals separate us by light years from creatures, such as chimps, with amazing intellectual capacity, enormous creativity, reasoning powers, a capacity for mathematical equation and an awareness, not only of immediate surroundings, but the entire universe *and* its origins. Such traits being seen in any way as a bequest from any preceding grunting simian type creatures seem too preposterous to contemplate, particularly as *every single* preceding entity lacked even the basic instincts to assure survival and proliferation, let alone bequeathing the superior human qualities to the final entity in the process, that is, our only convincing (Cro-Magnon ancestors).

It is clear, in paleo anthropological studies, that no single progressive evolutionary line or family tree can be traced back to any specific ancestor among the various varieties of Hominid. If Pro-Consul did ever exist, it would seem that the divergence along the ape line, rather than the proto-human line, was more successful, as is evident by their proliferation and continuous existence today, rather than being earmarked for extinction shortly after their appearance, as in the various Hominid types, including the appearance and rapid disappearance of Neanderthal, and of course the Homo Erectus creatures.

The period of existence of all the former entities becomes progressively *shorter* when surely it should be the other way around in an alleged 'naturally selected', and obviously approved process, and this can be seen by looking at the period of appearance to disappearance in (1) the Hominids, (2) Homo-

Erectus and (3) Neanderthalensis Sapiens, and brings us to a rather disturbing conclusion that Homo Sapiens by these massively reducing figures, are due to disappear any time now, but this is unlikely to happen, with the human having such advantageous features where such mental attributes allow the development of medical science than can prolong and indeed increase the chances of a longer lifespan, which will remain in an upward mode with further advances in genetic science and medicine.

To recall paleoanthropologists continual quest for fame and notoriety, Don Johansen and Blake Edgar [4] seem to agree when they said, "The pursuit of origins can be enjoyed in other ways than sweating it out in a tent in hostile climates, and that is the front page coverage in newspapers and magazines and best selling issues".... (in regard to possible discoveries). Clearly, this would be a suitable reward for those doing such valuable work in pursuit of their science today, who continue to 'sweat it out' in pursuit of the 'missing links'.

There is a strange paradox evident in books regarding the life and behavioural patterns of the Neanderthals in that anthropologists say there are indications that they buried their dead, and others say that the studies of their remains from certain sites seems to indicate a certain proclivity among them for cannibalism. In other words, an apparent *respect* for their dead on the one hand, yet this strange proclivity for *eating* them on the other. In any event, it would appear that however, we look at the Neanderthals they looked like animals, behaved like animals, lived a brute existence and cultural evolution seemed to play no part in their behaviour. They were unrelated to Cro-Magnon and it is stated did not inter-breed with him and quickly bowed out after Cro-Magnons appearance, almost as though it was purposely intended.

On the other hand, cultural evolution among the Aurignacion type was very evident and appeared to start around the upper Paleolithic period of some forty thousand years ago, and improved very rapidly with amazing artwork and created minutely carved artifacts.

As previously mentioned, attempts were made to link the ancient species to modern humans by studies and research into mitochondrial DNA only inherited from the mother and theories of a single Earth Mother, or 'Eve', began to emerge, but it was said, "These studies were shown to be statistically flawed in 1992" [5] . However, it cannot be avoided that if mitochondrial DNA is only inherited from the female then it must be traceable back to the formerly named single Earth Mother.

Richard Leakey says, "All the samples of mitochondrial DNA analysed so far from living human populations, are remarkably similar to one another,

Indicating a common *recent Origin(6)*. Therefore, that avenue of research came to rather an abrupt end. However, the term 'recent' could be applied to a period of forty to fifty thousand years ago in evolutionary timescale.

Richard Leakey, also says "... If genetic mixing between modern and archaic Sapiens *had* occurred, some people would have mitochondrial DNA a very different from the average, indicating its ancient origin. So far, with more than 4,000 people from around the world tested, *no such material evidence has been found*". Perhaps the expression 'archaic sapiens' should be 'archaic primates', as the Aurignacion, Cro-Magnon peoples would have had the same DNA as ourselves.

To many people, this would seem to very clearly indicate that our immediate ancestors, the Cro-Magnon peoples, who did mysteriously and rapidly appear on the world scene, had nothing in common at all with any alleged ancient ancestors, as said, certainly did not interbreed with their immediate predecessors, leaving us with the so far unaddressed question, 'Where *did* Cro-Magnon come from?' and this factor, certainly encourages the speculation of the possible 'third alternative' for human origins regarding some hypothetical ET experiment to advance human intelligence.

This very wide timescale, for the appearance of the still searched for Pro-Consul entity, or common apelike ancestor, is speculated upon as being anything from fifty million descending to around five million years ago, but the lower the figure becomes, the more it compounds the mystery, simply due to the fact that more recent bones ought to be easier to find than older ones.

It is quite evident from a simple visit to a natural history museum, that *much* older bones than those of even fifty millions years ago *have* been found regarding other earthly creatures. They must represent a frustrating experience to those working so hard to finalise the human question to see such creatures standing with every bone in place.

Paleoanthropologists, such as Don Johansen, have said, "Paleoanthropology faces contention from within and from without...." [7] and regarding the varying opinions of his contemporaries ....."Controversy is rampant among them". [8] The more eminent among them will readily admit to all, this controversy and disagreement that exists within this science.

After the trauma and shock that the theories introduced by Charles Darwin and Alfred R Wallace, began to subside a little, many people began to take stock of their beliefs, and made adjustments and concessions to science, while still attempting to maintain their religious convictions. They must have viewed it as simply adding insult to injury when being coldly told by science that,

"Creationism is mystic and not scientific ...." "Evolution is a fact....", and so forth, especially when the science of evolutionary theory did not explain, nor has it since explained, the Cambrian explosion of life forms, which certainly conform more to 'mystic creationism' than science, where the primary classification of marine life is still evident today, appear 'suddenly' in the sedimentary record, and also have had little success in constructing human effigies from ancient periods with every bone in place and especially creatures appearing with skeletal frames that appeared in the Cambrian explosion, with no fossil history to depict their gradual evolutionary formation such creatures would surely have required such a history to reach their respective bodily form classification. Nor can science explain the massive creation event of fifteen to twenty billion years ago in all its intricacies, which they had called the 'Big Bang'. There is a yawning gap, that cries out for explanation as to how this massive jump from primitive low life algae and single celled organisms could bridge it to produce these fully formed classifications of the Cambrian period yet leave no fossil history on the way, and these could really be termed the Pre-Cambrian missing links'.

This great biological discrepancy has as said, not been explained by science. Why should the process of natural selection in evolution of the human, be so dramatically different with regard to the major genetic alterations previously mentioned, to develop the many unique characteristics evident in the human? In particular, the amazing, mystifying, creative, and over endowed human brain, an organ that made Charles Darwin almost abandon his theory and his counterpart divert from it completely.

The question of this over endowment of brain material has not been solved and seems to give rise to question whether the time scales we apply to its development are correct or even possible. It we consider that a period of fifty million years has been necessary, that is, the time stated to be the onset of primate evolution then our alleged cousins have required that amount of time to develop their mass of brain cells or neurons, and it is said that human possess ten times as many, then logically, one could suppose that the human should have required 10 times as long to have developed theirs. This amounts to five hundred million years, that might suggest, that for a large proportion of time the human brain was developing elsewhere than Earth in our possible ET 'creators'. To illustrate the rapidity in which a small fragment of bone can quickly become bestowed with all the qualities of an alleged upright ancestor, more from wishful thinking than fact, is an event of 1932 in India where even a graduate student, with no experience yet amassed in the field, could create a new human genus, providing he could enlist someone notable to agree with his find, namely Louis Leakey. From a jaw fragment, a new human genus was created called

'Ramapithecus' from fourteen million years ago. Considering that date, how was the age correctly dated?

In 1935 the German paleoanthropologist, von Koenigswald, created an alleged 'ancestor' from a single tooth that he bought in a Chinese junk shop and called it 'Gigantopithecus', or giant ape. Louis Leakey found even older bones that dated back some twenty million years, and called it 'Kenyapithecus'. He thought it might be the elusive Pro-Consul, but few other authorities went along with him.

It is clear that paleoanthropology has been the most argumentative of all the sciences since its inception. If we listen to some of the comments from a cross section of people involved in the field, this is amply borne out. Professor Sherwood L Washburn of The University of California at Berkeley said, "The study of human evolution is a game rather than a science in the usual sense". [9] Fossil hunter, F Clark Howell, said "Anyone who feels that we have the problem (human origins) solved, is surely deluding themselves". [10]

Don Johansen, the discoverer of 'Lucy', clearly admitted, there is no such thing as a clearly defined human family tree going back to our alleged apelike ancestors, and says "The variety of human family trees now cluttering up the literature, makes it virtually impossible to identify the correct tree because of the forest. [11] He also goes on to say, "We know that some fossil groups like the robust Australopithecine *were not on a direct line of descent to modern humans.*" [12] To many, this would seem to suggest that they have no place in the fossil record as alleged quasi-human predecessors. What else can the statement, 'Not on a direct line of decent' mean except that they were not our ancestors and, of course, many would take the view that *none* of the apelike Hominids were on a direct line of descent to modern humans, but were just simply another species of slightly more refined ape, borne out by mitochondrial DNA studies.

There would not appear to be any valid reason for the complete disappearance of the entire range of six or seven Hominid types (probably not inter-breeding) from the face of the Earth, leaving the true pongid apes (without the alleged advantages of the Hominid), to freely proliferate. If they were not superior to the pongid apes then all these entities could be viewed as a mutational failure and doomed to oblivion from the outset, which would simply classify them as a rather unsuccessful but additional species of ape (or faulty mutation).

So what majestic hand wiped out this confusing array of Hominids that make the family tree look more like a bush? The paleoanthropologists, even though skull portions and scraps are all that have been found (except for the

'Turkana Boy' and the contentious 'Lucy') have all been busily getting the Hominids up on two legs and industriously making tools right from the early days through to those currently active in the field. The anomaly of 'tools, and no bones', and 'bones, but no tools', has still not been resolved. Microscopic analysis of alleged tools can soon reveal if they have been 'worked' or not, and therefore many stones that adorn the fossil table as alleged tools that may not show this evidence of 'working' could have been naturally formed by Earth's forces, such as in the natural flaking and separation that occurs in slate. Some anthropologists have accused their contemporaries of being too ready to assign possibly naturally formed stones to the 'tool' category.

The forty percent complete skeleton of 'Lucy', another alleged ancestor, has an apelike skull, was barrel chested with short legs, long arms, curved fingers and toes, yet, because of the paleoanthropological opinion regarding the pelvic region, up she comes on two legs with full bipedal locomotion, which must have been rather painful for 'her' with such a foot shape. The 'Laetoli' tracks found in solidified lava of alleged upright walkers were assessed as being some three million years old, similar age of 'Lucy', but other anthropologists have said, 'Lucy did not make the tracks', and also that 'she' might be a 'he', highlighting once again the obvious disagreement in all the finds and discoveries pertaining to our alleged ancient ancestors.

The 'Laetoli' tracks found in solidified lava of alleged upright walkers were assessed as being some three million years old, similar age of 'Lucy', but other anthropologists have said, 'Lucy did not make the tracks', and also that 'she' might be a 'he', highlighting once again the obvious disagreement in all the finds and discoveries pertaining to our alleged ancient ancestors.

Whatever other features of the human body form and its development and evolution that are discussed and argued over, the single most amazing organ is the awesome human brain. Not one single explanation for the rapid development of this amazing collection of cells, equal in number to all the stars in the galaxy, has ever been offered. Consider Isaac Newton poring over his calculations, or Albert Einstein working out his equations, then watch a chimp splitting open a banana simply because he hasn't the intellect to even consider how to peel it and try, if you will, to see any possible connection between them. It is really not surprising that many cling to their pre-Darwinian faith, or even contemplate the infusion of intelligent genes (which we ourselves are now isolating), by a superior Intelligence, 'not of this world', as proposed in an alternative theory for human appearance on Earth. They may say, "Well why not?" They must be God's creations as well; therefore, God (indirectly) would still be responsible for human creation.

Elitist scientists, who refute the latter assumption out of hand as sheer self-aggrandisement, or even science fiction, are simply putting blinkers on or closing their minds and stating, 'No science exists anywhere else in the Universe, and if it does, it is *all* inferior, or less advanced, than our own'. To many people such closed-minded statements would seem to be a lot more ridiculous than any alternatives to the 'apes and men' theory for human origins.

To consider that the human entity developed in a natural evolutionary way from the primates seems just as 'mystic and unscientific' as the Biblical version is seen by those who accept the 'apes to men' theory.

The strange fact is, that many who slavishly adhere to the entire 'Darwinian package' as Richard Leakey terms it, will avidly watch man-made artifacts soft land on Venus, trundle about taking photographs on Mars and flying around every planet in the solar system, then refute out of hand, any suggestion that God (or even 'the gods) might be responsible for it. Instead, they will favour a theory that such gifts are a bequest from ancient apes that are still unproved to be actual human ancestors and only possessed a small amount of brain cells.

They refer to chimps as 'cousins' seen jumping about in our zoos, that possess even less brain material and who have never created *anything* in all the fifty million years of their evolution. The subject of this profound blossoming of the human brain and its fantastic capabilities, in respect of the many human geniuses, confound all anthropologists and make them either change the subject or refer to the various traits observed in the chimpanzees while 'ape watching' and training and conditioning them, subconsciously forcing them to confirm to human-like activity that they wish was there naturally in the first place.

Even from the initial introduction of the theory for humans having simian ancestry, there have been misgivings from the start. From Alfred Russell Wallace up to the present day, many people have assumed some unknown factor rather than apes being responsible for this profound human intelligence.

With regard to those 'conditioning centers', chimps are not only simply observed, but trained and encouraged to do things that they would not naturally do in the wild, simply to allow the anthropologists to make comparisons to humans in regard to their behaviour patterns. This must be entirely due to their possibly coincidental, although vague, resemblance to the human body form. If this were not the case the same observation and conditioning might be equally applied and, perhaps, more easily achieved, with seals balancing balls on their nose, 'clapping', or rolling over to order, or the amazing intelligence of dolphins. Killer whales, elephants, dogs and many different species can be trained to do things they would ordinarily not do. A sheep dog, acting like a human herdsman from nothing more than a series of whistles, displays more intelligence than the average chimp, but because of their body form, anthropologists would not waste their time on them.

Horses are also very intelligent creatures and respond to training and conditioning, but clearly, so much time is spent with the chimps that there is an apparent compulsion to turn them into some kind of quasi-human entity. A factor little mentioned in the discovery of ancient bones is the large variety of animal bones often discovered with them. Not long after Darwin's explosive theory, the ardent searching was in full swing worldwide.

In 1987, a Dutchman called Eugene Dubois, found a piece of cranium in Java (along with a rich store of *other* bones) that even he thought was too low and flat to be from a human-like creature. When he returned the following season, he found a thighbone and immediately pronounced it as an ancient ancestor, *and* from the same skeleton of his previous find. From these two portions, he made the breathtaking assumptions that not only was it an ancient human ancestor that walked upright, but was also the elusive 'Missing Link'.

Here again, was evidence of this strong compulsion to make wide-sweeping statements regarding their finds, and the worry and concern evident in not finding this vital evidence to the point of every scrap of old bone being put up as a candidate for a primate ancestor. Not surprisingly, his contemporaries did not all immediately agree with him on his interpretation of his discoveries. In consequence, he rather petulantly locked them away for twenty-eight years.

Various missing link claims have been made ever since the Darwinian/Russell theory first emerged and all of them have subsequently turned out to be more in the line of wishful thinking than items of significance. Raymond Dart's discovery in 1924 of the so-called 'Taung Child' in South Africa was immediately pronounced as an ancient human ancestor, but others disagreed and offered alternative suggestions, such as that it was the distorted skull of a young chimpanzee, or that it was too young and underdeveloped in any case to assume what it would have looked like at maturity, once again highlighting the dissent, contention and disagreement apparent in the field.

Then there is the case of certain Dr Robert Broom, a Scottish fossil hunter of the same period and also a friend of Raymond Dart. He made it quite clear that a certain exploitation of the urge to find an classify old bones as possible ancient ancestors, or even missing links, did exist at the time, and it was clear that many people saw possibilities of financial gain if they could find bones before the professional searchers, desperate to find their 'Missing Links', and sell them to them.

Dr Robert Broom knew the quarry manager who had worked at 'Taung', the place of Raymond Dart's earlier discovery, and this chap was called G W Barlow, who admitted to Dr Broom that he quite often sold any 'nice old bones' that he found to certain visitors. Predictably, he then produced for Broom's 'a beautiful fossil brain cast only blasted out that very morning'. [13] Dr David Broom obviously found G W Barlow was a useful contact as a couple of years later Barlow handed Broom an 'ape man's' upper jaw that he had obtained *from a schoolboy.* Broom proceeded to hunt down and trace the boy, who promptly pulled four teeth out of his pocket, and the end result was that all these fragments were enthusiastically put together to produce yet another alleged ancestral genus 'Paranthropus Robustus'.

A photograph was depicted in *Time Life* in an article called 'The ancestry of Man', that showed three jaw bones, one of a Hominid and alleged pre-human ancestor 'Australopithecus Robustus', one of an orang-utan and one of a human jaw. The first two are very clearly apelike, with a robust strong looking structure and teeth in comparison to obvious difference in the human jaw. This glaring difference can be seen in all the bone fossils of skulls and jaws that exist, allegedly belonging to ancient human ancestors.

In the artistic reconstructions as to what these entities really looked like, they are always drawn as obviously human-looking, with a few exaggerated traits, enough to hint at apelike features, but retaining the distinct human image where, in reality, if we were able to behold an actual being, we would probably see very little difference in their appearance to a kind of apelike creature.

When one considers the large amount of plaster in skull reconstructions that, with all the missing fragments has, of necessity, had to be used, together with the obvious conclusion that those involved in the reconstructing, particularly in the padding out and attempts to make models of what they think their actual appearance was; there must exist an enormous, perhaps subconscious desire, to make the final product portray an image as human looking as possible.

To return to the 'stone tools' problem, or lack of them in the areas where the alleged ancient bone fossils were found, this certainly seems mysterious if these areas were quarry or 'tool' manufacturing areas. After all, the workers in them would surely have lived and died there, or at least in close proximity to them and once having made the tools, they would surely be using them in the said habitat, in preparing meals and so forth, so logically, there should be *some* tools evident with their bone fossil remains.

One could rummage among the scree at the base of almost any cliff face and immediately find at least half a dozen stones that could be said to appear to have been fashioned or worked to produce them, and the more recent and obvious stone tools used by our more recent and obvious ancestors, the Cro-Magnons, do actually show signs revealed by microscopic analysis of being 'worked', and no doubt some older ones as well.

The Time Life article on human origins stated, "Until much more evidence is dug up, the question of who begat whom among all the possible predecessors

of modern man is highly debatable and the provable record of man, and 'near man' is still obscure.

The full story of the Neanderthals is hazy and uncertain and it has been said that definite signs of retrogression have been noticed in the strange fact that more recent finds seem to indicate more simian traits than earlier bone fossil discoveries. With regard to our more obvious ancestors, the Magdalenian or Paleolithic men, an even deeper mystery surrounds their very sudden appearance. There have been cave excavations depicting Cro-Magnon bones in the strata lying immediately above those of Neanderthal with no evidence of intermingling but rather signs that Cro-Magnon simply took over and replaced Neanderthal almost immediately in comparison to normal evolutionary time scales.

Although anthropologists have been urged, during their eager campaign, to confirm the apes to men theory, to withhold early judgment and avoid basing broad and far-reaching theories around their fragmentary remains, the previously mentioned evidence simply speaks for itself. The Cro-Magnon came, the Neanderthals went, that much is obvious. What is less obvious is how and why and for what reason?

There is no good reason why we should not be able to find all the necessary bone fossil, even complete skeletons of alleged ancient ancestors, as bone material seems almost indestructible when we consider that no plaster is required to assemble all those dinosaur effigies. Every single bone, in most cases, is in place and there to see, and they have been dead for over sixty five million years. Far longer than the oldest Hominid skeleton of Australopithecus Afarensis, yet all we ever seem to discover are scraps, fragments of bones, an odd piece of skull or jaw and some teeth. On average, we must be overdue for another profound discovery of a complete, or almost complete, skeleton coming to light somewhere around the globe.

According to the *National Geographic* of February 1997, one of the greatest mysteries of paleoanthropology is when, where and how 'homo' replaced the Australopithecus genus. They were apes in all respects, except for the alleged bipedalism and, as said, considering the large amount of varieties in the primate species, there is no good reason why they could not have been just another. In any event, the bipedalism did not seem to assist in preventing their extinction along with all the other Hominids in the family bush said to be around six or seven in number.

As previously said, although links have been attempted to be made with the alleged upright walker 'Lucy', (Australopithecus) and the Laetoli tracks, where human-like footprints were apparent in solidified volcanic ash and dated a three million years old (the same as Lucy), Russell Tuttle of the University of Chicago argues that the curved toes of Lucy, who would be happier in the trees, did not make the tracks.

Peter Schid and Martin Hausler of the University of Zurich, after studying the pelvic of the A Afarensis, did not seem convinced that Lucy was female at all and suggested 'she' might be a 'he'.

Some anthropologists have the Hominids lighting fires. Others say that the fires were probably natural occurrences. It is difficult to see how this assumption that all the Hominid was bipedal can be verified from the scraps of their bones we possess. Don Johansen admits that "An individual Hominid genus has been represented by anything from a single tooth to an almost complete skeleton", [14] and he has admitted to the existence of unresolved questions. He said, "One of the key unanswered questions today is why a modern anatomy has preceded modern behaviour". [15] This clearly indicates that in his view a more developed brain should have encouraged bipedalism, rather than the other way around, which would be the case *only if* the Hominids *were* bipedal.

Don Johansen also says, "We cannot answer exactly why we evolved our large brains". [16] Perhaps the word should have been 'over-endowed' rather than 'large' since size itself is irrelevant when we consider, for example, the twenty-two pound brain of a whale or the large brain of an elephant. What seems to be more important is the number of actual neurons, or brain cells, active within the tissue and the equally important point of size to weight ratio.

Therefore, the biggest mystery is slow, plodding impersonal evolution and natural selection could have been responsible for the amazing creativity, intellect and profound qualities of the human brain, when it did not see fit to suitably over endow any other species on Earth so that we could talk to the dolphins or play chess with the chimpanzees. Only the human has been singularly over-endowed and selected for these capabilities within the area of the brain that governs the higher functions.

In the spring of 1998, a series of programmes dealing with evolution where shown on BBC 2. One would have thought that the unanswered questions would have been dealt with and some attempt made to address them.

However, this hard shell of apparent unassailability seems to make people in general afraid to question anything to do with it and anthropologists have this air of elitist specialists, who have all the answers, when in point of fact quite the opposite is the case. They cannot explain the more profound points and rarely agree with each other.

During the afore-mentioned programmes, Melvin Bragg made some attempt to address the point of intellectual development in humans compared to apes and, sure enough, a lady anthropologist began quoting the behaviour patterns observed in the ape watching colonies. Amazingly, she interpreted the observation of a chimp, who ambled down to the water's edge, did not drink or eat, but just looked, then returned to its point of departure, as a profound revelation of abstract awareness of the beauty of nature and its natural habitat. In other words, the higher qualities of humans had, at last, displayed themselves in chimps. It did not seem to occur to her that it might have been absent-mindedness, pure and simple.

If humans (who are light years away from chimps in terms of intellect and the possession of high qualities) can be absent-minded, then it is a pretty safe bet that apes will be too. This gave rise to such phrases as 'the absent-minded professor'. People of high intellect with their minds constantly occupied with problems they are trying to solve can be so pre-occupied, that when doing everyday mundane chores, can be forgetful, such as going into another room and forgetting why. This seems to indicate that perhaps too much ape watching is leading to too many false assumptions on perfectly natural and unspectacular behaviour, and this tends to make them a little too eager to assign human qualities to them. It is clear though, that at least some scholars and academics open their minds to other possibilities and are not afraid of 'professional suicide' by mentioning them. We could refer to our earlier comment where academically qualified people such as Ralph Franklin Walworth, author and member of the U.C.L.A, said "Although we find vast deposits of the fossilised remains of now extinct animals, remains of our ancient ancestors are extremely rare and are of such rarity that the finding of a single fragment of bone can cause worldwide headlines" [17] and, more profoundly, "Sub-humans underwent a metamorphosis and become modern people. Perhaps, as certain people have suggested, *visitors from another planet caused the change"*. Moreover, "Perhaps the biblical account in Genesis *is* correct. Humans become larger and stronger, spines changed their shape enabling them to hold their heads erect. The oral cavity and teeth changed, allowing more room for movement of the tongue for speech to develop and the brain case of the skull to grow tremendously. 'Suddenly', people developed the intellectual capacity to create great cities and civilisations and the means by which this happened remains mysterious. Of course, it only remains mysterious until we contemplate on A R Wallace's 'unknown variable' and what it could be.

What then was this process? What force motivated it and singled out humans for these amazing changes from alleged 'brute hood' to amazing intellects and great creative achievements in the arts of sciences? It is clear, or should be, to those avidly watching the apes, that in all their immensely long evolution they have never done anything they were not taught to do, or used their hands for anything other than simple tasks relative to feeding and their everyday existence. Even bipedalism and manual dexterity would not guarantee any human capability without the excess intellect being *in position in the first place*. In other words, the brain motivates and utilises the body form and not the other way around. Otherwise, after forty to fifty million years of evolution, the chimps should be teaching humans in *their* universities and would have conquered interstellar travel long ago.

Human mental evolution has only made itself markedly apparent in *thousands* of years, not millions. Although attempts are made to push back the time for the first appearance of alleged proto-human, the further back we go, the cloudier it gets, even the Hominids appear questionable as the progenitors of humanity. Home Erectus appeared six hundred thousand years ago, and still had an ape-like skull. Major brain development only seemed to occur in the Neanderthals/Cro-Magnon transition of thousands of years ago which vastly compounds the questions as to how this 'miracle' happened.

Apes can be trained to be self aware. If a mirror is left in their cage for long enough they will gradually get the message that they are seeing themselves, but the day a chimp does something 'naturally' that appeared quite profound and that they were not trained to do, such as carefully peeling a banana instead of splitting it open and scooping out the fruit, or attempting to form words, or copy human speech, or maybe count their fingers, that will be the day that many people will sit up and take notice. There are those that will say chimps are a separate line of divergence, or 'modern apes', that branched off from the elusive Pro-Consul and should only be considered as a distant cousin, and should not be expected to display human qualities. Well, if that is the case, then what is the point of all the chimp watching in the first place? In addition, why the eager reports when the watchers think they have observed something profound in their behaviour, such as the afore-mentioned, and quite probably, absent-mindedness of the chimp when shambling down the river to just stare? In addition, the intensive 'training' to make them do things they would not naturally do, so people can think they are naturally displaying human-like qualities who may not be aware of the 'training'.

The profound gift of creative intelligence does not only exist in the average human brain, but in a single bodily cell, the human blue print for another complete body form. Human DNA is a chain of molecules. Molecules are being detected in space. Does even basic amino acid have intelligence? How does it know that it needs the protection of the cell membrane?

Amino acids arrive from space in meteorites. The Murchison meteorite that landed in Australia contained seventeen amino acids, some of which were alien to Earth.

With regard to the afore-mentioned left and right-handed configurations we are told that no right handed amino acids exist naturally on Earth, only left handed. The Murchison meteorite had five left handed and twelve right-handed amino acids. Scientists pick up meteorites from the polar wastes. Not only are they easier to spot, but there would be obviously less likelihood of bacterial contamination in such zones. When we consider the amount of celestial rocks that reach the Earth, it is surprising that with all the cities and built up areas, that there is hardly any 'collateral damage' events.

Amazing statements have been made after analysis of meteorites, such as "Wherever this rock originated something once lived". [18] The 'Panspermia' theory for the origin of life would only be possible if the host planet was suitable in the first place and the Earth environment, with its favourable position in the habitability zone of a stable star, is exceptionally favourable, like a fertile womb receiving the life seeds. There would be little chance of meteors, ripe with amino acids, raining down on Mercury, for example, ever achieving a life process initiation. It has been suggested that certain bacteria has arrived from space in the past such as influenza germs and some correlation has been made with cometary visits and this raining down of bacteria into Earth's atmosphere as the Earth passes through the long cometary tail.

The Earth is literally infested with life that permeates everything in contrast to the hostile barren or gaseous worlds that make up the rest of the system and with regard to the amazingly rapid development of humanity, there could not be better environment for a being such as the human to evolve, but that goes for all the other creatures on Earth as well, but they all obeyed natures laws in their development that is extremely slow evolution and natural selection, but humans did not. They seem to have arrived on the scene with the appearance of Aurignacion mans – *already evolved.*

If we take the Neanderthals out of the equation, which seems natural if, as has been stated, they were unrelated and non-interbreeding with Cro-Magnon then it compounds the problem even further. The Neanderthals were a mystery all by themselves. It took 400 pages in a book for the writer to conclude, "They remain the subject of debate", and "The puzzle remains undetermined" [19], etc.

In spite of Africa being the popular venue for the origin of humanity, a work on human evolution stated, "Thus far, absolutely no fossil record has been uncovered in Africa that would indicate the immediate ancestors of our nearest living 'relatives', the chimps and gorillas". [20] The same book confirmed the afore-mentioned burst of cultural evolution apparent in fairly recent times, with "A radical and abrupt change in the archaeological record occurred forty thousand years ago, termed the creative explosion". [20]

It is said that no 'Homo Erectus' bone fossils have been found in Europe, only Neanderthal and Cro-Magnon. Home Erectus is alleged to be a bipedal upright tool maker, with a brain of some 900 cc compared to modern man's capacity of some 1,350 – 1,400 cc. his skull, however, is said to be more pithecoid than human and, with the scarcity of bone fossils quite a lot of conjecture has been necessary with regard to the attributes and activities of 'Homo Erectus'. It would be understandable for there to be no bone fossil found on island landmasses, such as Great Britain, but Europe would eventually have been reachable if Homo Erectus *was* possessed with the qualities certain anthropologists have bestowed upon him. Surely, he could have forded the narrowest part of the Nile, if he could cross the pacific.

In certain television programmes, [22] amazing conclusions were made about the possible achievements of Homo Erectus from an alleged Australian stone tool dated as *two million* years old. In other words, long before the aborigines arrived there. The age of the artifact was determined by the thermo luminescence, or T.L.P. process, which immediately cast doubt on the accuracy of the process itself.

The usual follow on assumptions then began to be made without any bone fossil evidence to hand. They resulted in the status of Homo Erectus being elevated to intrepid explorer and bold adventurer, accomplished boat builder, navigation and explorer, without a single bone in sight.

We must, therefore, agree with Richard Leakey when he said, "Until many more relics of human pre-history have been unearthed and analysed, no anthropologist can stand up and declare, this is how it was in every detail". [23] The aforementioned quote from the book *Making Silent Stones Speak* was fairly pedantic regarding Africa as not being the place of origins of our alleged ancestors, this contradicts Darwin's viewpoint.

Richard Leakey quotes a passage from Darwin's work in his book *The Origins of Humankind,* part of which said….." It is therefore probable that Africa was formerly inhabited by extinct apes closely allied to the gorilla and chimpanzee and, as these two species *are now man's nearest allies...*" [24] This was a breathtaking assumption for Darwin to make in his time, when not a single bone alleged to be from an ancient human ancestor had even been found.

It would appear that it was Charles Darwin who set the pace by example for future anthropologists to make equally breathtaking assumptions regarding the scraps of bone and teeth that they subsequently found and indeed chimpanzees and gorillas are still referred to as 'ancestors' and 'cousins'.

Richard Leakey bears out the above in his book by qualifying Darwin's presumptuous statement with the following: - "We have to remember that when Darwin wrote these words no early human fossil evidence had been found anywhere. His conclusion was based *entirely on theory*. [25] It makes one wonder how many significant fossils may have been uncovered during building and excavating that was obviously occurring during Darwin's life and simply discarded by not being recognised for what they were. Of course, anatomically modern bones of the Cro-Magnon peoples had been found in Darwin's time and those of the more apelike Neanderthals, but the early pathologists and paleoanthropologists were making some rather odd and particularly disparaging remarks about the latter in those times. When the shock of the Darwinian pronouncements rippled through the British Establishment, it was disturbed almost as much as the Church was by these challenging revelations. The British Establishment, however, was more disturbed about Africa being singled out as the birthplace of the 'noble human', after all, with the domineering colonial attitude that prevailed at the time, in England at any rate, Africa was seen as the birthplace of the 'savages' and the so called 'fuzzy wuzzies'. But that kind or arrogance was also displayed in other lands when we consider the slave trade, which Britain played a major part in putting a stop to such an odious practice?

Charles Darwin appeared to take a simplistic and rather logical view, that once ancient man (or the Hominids) stood upright, this would naturally free the hands for tool making and other enterprises that would ensure the gradual development of intelligence. The fact is, it did not happen, and all the Hominids died out after making (if we can trust our dating process) only the crudest stone tools. However, even with these basic achievements, they were surely more advanced than their true ape contemporaries were. Yet the true apes survive while the Hominids all meet their demise. Why would evolution provide other creatures with hand-like extremities and give them bipedal stance, yet hold back on one vital factor that would enable them to utilise such advantages, i.e. *intellect?*

Tyrannosaurus Rex, we are told, lasted for *twenty million years*, thirty-five times as long as the period from Homo Erectus, to modern man, yet only used his bipedal and arm-like appendages and claws to savage and rip open other animals in his carnivorous attacks. The kangaroo is another example, not to mention all the primates who have never exploited their advantages for anything more adventurous than their survival routine. This was through no fault of their own. They simply had not been provided with the intellect so to do. We may also consider the bipedal ostrich, when it lost the ability to fly (assuming it ever had it) there is no sign of its stubby wings morphing into limbs.

It is clear, therefore, that Darwin's simple, but logical assumption at the time was incorrect. If it *was* correct then we would be surrounded by ape scientists, ape physicists, astrophysicists, and apes sitting at drawing boards designing spacecraft. After all, they have had an enormous period of time up to the evolution of modern man since their evolution began from the early primates.

In spite of chimpanzees often being referred to as our 'cousins', the blood proteins of chimps gorillas and humans are quite different from each other. Richard Leakey says, "According to conventional wisdom, chimpanzees and gorillas are each other's closet relatives, with humans standing a great distance apart"... [26]

With regard to the earlier claims concerning a creature singled out for a new human genus and called Ramapithecus, that later research stated was *not* a candidate for human ancestry, going back fourteen million years, (because of human failings in wrongly assembling the fossils) signifies the over-exuberance, even in famous forerunners, in the field that did not recognise the error, such as Louis Leakey, who not only gave his blessing to 'Ramapithecus', but saw an even earlier human ancestor in his 'Kenyapithecus' of twenty million years ago, which was also disputed.

His son Richard, however, seems less prone to impulsive, wishful thinking and reserves judgment, borne out in his statements in his books, such as "The task of inferring an evolutionary link, based on extremely fragmentary evidence, is more difficult than most people realise and there are many traps for the unwary". [27]

The fact is, that the entire alleged ancient human fossil evidence is little more than fragmentary, yet many anthropologists, if not all, remain firmly convinced that human development from ape-like ancestors, is an established fact on the one hand, yet the more honest among them admit that they have not discovered the final proof on the other.

It is not inconceivable that the over-exuberance displayed in the past may not occur again in the discovery of a future bone fossil, and the prevailing anthropological opinions ensure that the same mistakes are not made all over again that were made in the past, as the same strong desire is still apparent to be the one to find it. As Richard Leakey put is ... "It exposed the folly of a slavish adherence to the Darwinian package". [28]

The confidence among anthropologists must have been more than a little shaken after they had excitedly constructed an entire behaviour pattern regarding 'Ramapithecus', only to have him stripped of his title of Hominid later and dropped as a candidate for an ancient human ancestor.

To quote Richard Leakey again, "Concrete evidence of the *inadequacy* of the Darwinian hypothesis is to be found in the archaeological record". [29] What he seemed to be referring to was the step by step logical assumptions of first coming down from the trees standing erect, thus freeing the hands, then finding manipulative creativity, and all this leading to the development of intelligence and tool making, but with the archaeological evidence confounding these logical steps.

The monumental change to bipedalism was extremely profound and deserved an adequate explanation as to how such modifications to bone structure in the hips and major changes and re-routing could have occurred so easily and so rapidly, when all around us and in every other earthly creature, genetic stability seems to rule, particularly in foliage, fishes and insects and millions and millions of years ago by without the slightest change.

A logical conclusion could easily be that some 'outside force' purposely brought about such changes. After all, humans have done exactly this themselves in dogs and cattle and horses, and the hybridisation of plants and genetically changed vegetables and so forth, and it seems clear that if the creatures and objects singled out for such manipulations were left alone, they would stay as they were. The only earthly creatures that have seemingly gone through all this rapid shape-shifting and genetic change, *and* in comparatively recent times, is *man and his recent ancestors.*

There is a glaring contradiction in one example of the widespread divergence of opinion among anthropologist. Don Johansen says, "Bipedalism did not make humans inevitable" [30], and Richard Leakey saying, "Without the bipedal adaption they (the Hominids) could not have become like us". [31]

The 'brain racking' that has gone on to attempt to account for this monumental change to bipedalism has also invited other various contradictions among the anthropologists. One suggestion offered was that some drastic geological change or catastrophe might have produced a recipe for biological change into proto humans, by *forcing* bipedalism on them by readapting to their changed environment, rather than moving to a new one, perhaps due to an unfordable river or mountain chain.

Maybe a huge fire destroys all the trees and the tree dwellers have to walk on the ground instead. Then, another suggestion comes up that far from them being troubled by geological barriers, proto-humans could become bold adventurers and even cross the Pacific to Australia, but leave no bone fossil evidence to confirm it. Certain anthropologists make such remarks as "The early humans of seven million years ago, etc. etc". Others will say, "From the earliest part of the fossil record of about four million years ago". Don Johansen claims that his discovery 'Lucy' was the forerunner of the Hominids (of three million years ago), yet says "The probable time of a common ancestor (Pro-Consul) for humans and African apes of six to eight million years ago .... etc. etc. [32] Others will speak of perhaps fifty million years for the initial appearance of this alleged common ancestor. It is really not surprising that the average layman is very confused with all this variation of opinion and chooses, for the most part, to opt out and simply assume that one day there might just be a consensus of opinion.

As said, at the beginning of the work, one of the most important aspects of geology archaeology and the discovery of fossils is to have accurate dating instruments and chemical analysis to achieve precise information.

Today, this compulsion to create a new human genus seems to have settled down and now some six to seven candidates as human ancestors seem to have established themselves. In the case of the Australopithecus Robustus entity, the skull is rather flat faced and grotesque with a large sagittal crest on the top of it, and it would seem to have looked little different from a gorilla whenever it lived. When looking at this skull, one can only conclude that no matter how much the skulls are padded out in the theoretical constructions and drawings to view them as possessing any human-like qualities, it is extremely difficult, if not impossible, to view them as such, yet the graphic artists are always encouraged in depictions of an ancient human entity to have him looking as human looking as possible.

The more gracile Australopithecus Africanus fare little better, rather like comparing a chimp to a gorilla. The robust Zinjanthropus Boisei, found the Olduvai Gorge in South Africa in 1959, is another grotesque skull assembled from many pieces and liberal amounts of plaster, and it is considered that the teeth point to human-like characteristics. Again, can we be assured that every single piece (amounting to hundreds) was patiently checked and dated? All are quickly brought up on two legs, even if only (as in the case of Homo Habilis) a few portions of skull are found. Homo Habilis was Louis Leakey's 'Man, the toolmaker', and two legged tool producing 'human' from a small portion of skull.

The accepted brain capacity in Louis Leakey's time for a creature to be assigned the title of 'homo' was 750^{cc} proposed by the late medical anthropologist Sir Arthur Keith. As they had deduced that this new skull fragment (found, incidentally, by Louis Leakey's son Jonathan) may not be able to qualify for a new genus and fit into the homo classification with its mere 650^{cc}, it was decided to 'move the goal posts', as it were, and Louis Leakey promptly decided to proposed changing the rules in order that his son's new find could be admitted to the family tree. Again, this compulsion to establish Darwin's theory as fact, by many means.

In *The Origin of Human Kind*, Richard Leakey states.... [33] The fossil record quickly becomes sparse beyond two million years and *blank* further back than four million years, yet frequently throughout the book refers to 'human' ancestors of seven million years ago. However, he does admit to the paucity of human fossil finds in the work by saying "These are slim pickings indeed upon which to recreate a picture of early human pre-history". [34] As it stands, there seems little point in postulating on any fossil older than three million years. Donald Johansen claims that all the Hominid finds after Lucy were 'her' descendants, yet other within his science not only consider Lucy as a 'he' but state that she could not have been bipedal with curved toes and more or less describe her as fitting the description of a chimpanzee with regard to the skeleton.

Studies of the various classifications of stone tools and who made them, is equally as contentious as the creatures of the fossil record. Some anthropologists state that only 'homo', that is, anatomically modern humans made stone tools and not the Hominids.

Certain anthropologists use the term 'different kinds' of humans. At the beginning of Chapter three in *Origins* Richard Leakey says, "We have come to see that 'Homo' (Sapiens?) was a different kind of human from its first appearance? [35] People might ask, how can we *have* different kinds of human? There is only one entity that, at this time, can be clearly viewed as Human and that is the Cro-Magnon, our immediate human ancestor. All the rest are little understood, argued over and contentious right back from (and including) Neanderthalensis Sapiens.

Biologists have stated (regarding the human helplessness at birth) that human offspring are 'born too early' and based on certain studies of the other primates that we are alleged to be related to, they state that instead of the current nine months in which the human mother carries her child, it should be more in the region of twenty one months and they cite this as the reason for the human child's absolute helplessness at birth, but how could human females give birth without surgery to a twenty one month old child? We may also postulate on the difficulties, that female Neanderthals must have had with their long skulls.

The question must now be asked, if humans are completely indigenous to this planet and are the direct result of an enormously long list of creatures gradually changing into something else, said to have occurred ever since the ancient lung fish left the primordial sea: would such evolutionary and natural selection processes, that so avidly take care of other species and worldly creatures, by having them on their feet and feeding totally unaided, be so careless with the young of the dominant and most intelligent species of all by having their young lie utterly helpless and vulnerable to extinction without constant care?

With the increase in brain size, and consequently larger skulls of the human, evolution does not seem to have caught up by enlarging the pelvic region in females, which must have been easier for Homo Erectus females compared to most other creatures, who lay, or give birth quite easily, and carry on with their daily routine very shortly afterwards. Human births are markedly more difficult and often result in Caesarean section and it is quite probable in the future, if we do have increase usage of brain material and possibly larger skulls, that *all* births may be restricted to Caesarean section. As said, Neanderthal females fared even worse but did not have the option of surgical help.

Births could not have been a very straightforward process among the Neanderthals with regard to their larger than average brains compared to modern humans. Their skulls were ape-like but longer from front to rear without the straight forehead of the human 'built' to accommodate the 'forebrain', the seat of higher human functions. Although the Neanderthals are still debated, argued over and their history unresolved, it seems that they were an evolutionary dead end and had little, if anything, in common with the Aurignacion or our Cro-Magnon predecessors, who arrived like the second showing in the cinema with the Neanderthals leaving after the first showing. We have said that the human defies the processes of natural selection but Neanderthal makes no sense of such processes at all by gradually becoming less evolutionary advantaged, evident in later fossil finds.

Although the Neanderthals were widespread throughout Europe and lands to the east, it is difficult to describe them as a successful creature, and this is borne out in their rapid and mysterious demise. Their robust, short limbed, stocky body frame could not really be attributed to colder ice age conditions as they existed in earthly locations said to have been unaffected by the ice sheets. In any case, they did not seem to exist long enough for nature or any natural evolutionary processes, to have had time to modify or equip them anatomically for existence in any particular earthly zone.

The Neanderthals were surely a subject where puzzlement appears to take over from logic, and they seem to fit the bill as an entity that 'went wrong', or were a mistake in some attempt to create the ideal humanlike body form and, as such, appear to have been quickly eliminated, or purposely retrogressed, perhaps by genetic means. Those subscribing to the fantastic 'ET biogenetic' process could see the short time of their existence, that is, in terms of long evolutionary processes, as fitting in with the proven history of the 'time dilation', where possible extra terrestrial biogenetic creators could streak away from Earth at super light speed with little time passing for them, as the continued on toward their next profound mission to introduce the amazing gift of intelligence elsewhere, but with a very large amount of time passing by on Earth, then returning from other cosmic operations to Earth to asses, modify or dispose of their genetic 'mistakes' or failures and improve where necessary.

What this all means, of course, is that these amazing postulations, that *still* seem more like science fiction rather than fact, could have occurred with each earthly mission, improving on the achievements and experiments of previous visitations to Earth. These postulations encourage the possibility that humans and their appearance reflects an 'experiment'.

Natural evolutionary processes appear to demand enormous periods of time even for the slightest change, but Neanderthalensis Sapiens was quite an enormous change in a comparatively short time from the Homo Erectus entity Neanderthal occupied a large swath of land across Europe, the Mediterranean areas and the Middle East, yet modern or Cro-Magnon men swept them away completely from Europe thirty-five thousand years ago without, it would seem, even laying a finger on them. Their appearance and demise depicts a failure, whichever way we look at it.

In the past excavations, where hundreds of tons of sediment are removed and sifted through to produce many bone fragments and patiently joined together to build up skulls and skeletons, can we be absolutely assured that *every* single piece has been accurately identified, dated and pronounced with one hundred percent certainty as being human and definitely all came from the *same* human skeleton?

Humans must have happily co-existed as they do today in certain areas of the world with apes and primitive groups living together. The bones of both human and apes would be, and have been disappearing into the Earth for centuries. Considering factors such as water erosion, perhaps earthquakes or geological upheaval and earthly predators scratching them out of the ground, pulling them about or dragging apes' corpses into caves that they have once been occupied by men, and so forth, surely this is a certain recipe for a great admixture of human, animal and apes' bones to have been fossilised together and increase the risk of misidentification of fragments as mentioned in the earlier quote from the British Museum press. If some of the fossils or portions of bone where dated exactly the same this would compound the risk of assuming they belonged to the same skeleton.

There must have been a higher potential for error in the earlier days when it seems necessary to prove the Darwinian theory as quickly as possible particularly as so many people had put their trust in it implicitly by abandoning their former beliefs. Is it possible that we might have to rethink the entire theory or rewrite the Hominid history if *every single piece* of bone making up these older skulls was re-assessed.

Anthropologists are attempting to push back the date of the appearance of modern humans as far as possible, which originally was little more than thirty-five thousand years with regard to the Cro-Magnon, or anatomically modern
of man, but this is viewed with consternation by certain anthropologists and the consternation and argument and debate among them all still continues as ever, but we could assume that a period of around one hundred thousand years ago saw the onset of modern anatomy.

Richard Leakey says, "Where does all this leave us? The important issue of the origin of modern humans remains unresolved despite the welter of information that has been brought to bear". [36]

Human intelligence is a power for great good or evil and past human activity has amply proved this in the advent of many saints to Satanists among us, and this strange equilibrium (in most of us) of *both* forces. Genetic studies are suspected, feared and rife with taboo when it comes to certain activity in the field, but striving to eliminate the negative forces of the mind would not appear to many to be questionable and, if discovering the means to do so, many would encourage its implementation, especially since we have discovered a gene responsible for intelligence there is no good reason why genes that are responsible for negative behaviour could not be discovered.

Returning briefly to the 'biogenetic creation' hypothesis, such creators would clearly have the capacity for such process. Postulating on how 'they' would, or could, go about it (with or without our co-operation) will be dealt with later, but we need not get too concerned about this, as the intelligence bestowed in the human brain, in the best examples, may well allow our geneticists to deal with the problem.

The development of the fantastic human brain and its amazing over-endowment of intelligence is a topic that should be widely debated among the anthropologists worldwide, but it is guaranteed to make them uncomfortable if dwelling on it for too long, or make them refer to the apes they are 'training' subconsciously in the colonies. Some will take the view that it was the advent and development of human communication, or language that caused the brain to evolve rapidly and increase its neuronal capacity. Others will say that it was the other way around. However, whichever way we look at it, the human brain development and its mysterious capacity and over-endowment is most certainly an enigma and cannot be taken for granted as a naturally evolved organ. Until it becomes clear, how this obvious defiance of natural evolutionary processes occurred, the mystery of the human brain will always be the subject of wonderment and speculation. Later we will discuss the amazing mysterious gene discovered, that allowed the development of the suitable larynx and that allowed speech to develop. Some anthropologists, as said, have assumed speech development occurred before the major development of the brain…….

But since the area of the brain said to be responsible for the gift of speech is in the forebrain and identified as the 'Brocas' area and the human skull had to 'shape-shift' rapidly by mysterious genetic processes to modify the skull shape to accommodate this additional brain material, that sets us so far apart from the apes, it would be more logical to assume that brain development occurred *first*, in order for the controlling factors to be in place for speech to develop. Otherwise, we would have to concede that a *different* area of the brain was initially responsible for speech, then the use of it evolves the brain, and then a different area (i.e. the Brocas area) develops and takes over the job. The latter would seem to be a more clumsy hypothesis. However, like everything else to do with the rapid and mysterious explosion of intelligent development, such factors will always be the subject of debate.

Paleoanthropologists seem to basically agree that Neanderthalensis Sapiens showed little sign for having possessed the ability for language by studies of their basic crania, or lower skull, yet they also say that the analysis of the basic cranium areas in archaic human remains of three hundred thousand to four hundred thousand years ago seems to show that they were equipped with the structure necessary for speech to have developed.

Richard Leakey says, "Within this evolutionary sequence we see an apparent paradox. Judging by their basic crania the Neanderthals had poorer verbal skills than other archaic Sapiens several hundred thousand years earlier. Basic cranial flexion in Neanderthal was less advanced even than in Homo Erectus". [37]

This seems to be another pointer towards the conclusion that the Neanderthals 'retrogressed' (naturally or otherwise). Other anthropologists have also pointed to these signs of apparent retrogression in the Neanderthal species, and it has been stated that finds dated as more recent than others have shown signs in the bone structure of being more ape-like and brutish than their ancestors, as well as degenerative bone deformities.

Everything seems to indicate some 'unknown factor' deciding that Neanderthal was a failure, then clearing the path, so to speak, by eliminating him and preparing the way for the grand entrance of the Cro-Magnon peoples.

Cro-Magnon is not so much judged by his excess of brain material as studies seem to indicate, rather strangely, that Neanderthals, in spite of his eventual demise and disappearance, actually had a larger brain, but it was clearly the arrangement and quality and positioning of the neuronal cellular materials that mattered, and Cro-Magnon's skull duly changed to accommodate this 'mystical and magical over-endowment' of intelligence that, at the time of the appearance of Cro-Magnon, was entirely unnecessary for everyday existence, but *was* necessary for speech, the arts and the higher pursuits of man to develop, and so began his cultural evolution, said to have occurred some thirty five to forty thousand years ago.

Those people involved in the patient observation of chimps in the African colonies or reserves, are also patiently 'conditioning' them and training them to become in their behaviour as humanlike as possible. If they had been left entirely alone, then visitors would not see anything different in their behaviour patterns than one would expect to see in the average city zoo, so one must ask what is the purpose of this 'training' programme in the first place. The main priority should only be to protect and care for them.

However, in one of these special areas one might be amazed to see a chimp knocking pieces of stone together to obtain a sharp slice of stone to cut the string of the box, so he could get at the hidden fruit. Hopefully then, the observer would go away feeling that what he had witnessed was clearly another indication that chimps are *really* our cousins. The chimp had been trained to do it and would not attempt to do it in the wild. These activities could be viewed as part of the desperate last hope to reinforce the apes to men theory in the light of the failure of the fossil record to do so.

Observers watching dolphins jumping through hoops and seals balancing balls on their noses would go away without retaining any feelings that such creatures might be our cousins, neither should they, when witnessing any tricks chimps have been taught. Chimps have 'utilised' certain things, such as twigs and sticks to poke and nests, but in all *their* massive evolutionary period, they have never, without human interference, created a single thing in their entire history, whereas other creatures with no chance of being regarded as 'cousins', such as birds, bees, otters, beavers, and so forth, most certainly have.

In another work I said that it seemed strange that with regard to the profound physiological differences in chimps, particularly their undeveloped brain and lack of certain areas that produce the unique human qualities, that the perpetrators of an earlier experiment chose to waste their time on a rather pointless exercise. The experiment took chimps out of their natural habitat at birth and introduced them to a human environment by treating them like human babies, with nappies, bottle-feeding and being spoken to every day.

The expectation seemed to be that the chimps would turn into a quasi-human child and exhibit human qualities, or perhaps start saying 'mama'. Surely, the perpetrators would have been aware that such an experiment was doomed from the start, assuming that they *were* cognisant of the fact that the physiological differences of the chimps, such as the high larynx and small undeveloped brains, would produce nothing other than experimental failure. Is this just another part of the desperation to 'make' chimps fit the descriptions of Darwin when he referred to them as relatives or cousins? Or an operation, viewed as having nothing to lose and perhaps something to gain? This desperation is, in a sense, quite understandable, when we consider the long period of time that the bone fossil evidence has been searched for, that is now beginning to indicate that such evidence may not be there in the ground to be *found.*

It is of little use concentrating on the qualities, and nurturing them, displayed in chimpanzees, then ignoring the amazing displays of intelligence in other species. Sheep dogs, for example, they do not seem to need to bark or act aggressively toward the sheep, but display their agility in rapid movements which clearly convey to the flock what they should do and where they should go, not panicking them but enforcing the canine 'will' upon them.

One can only imagine how the flock would react if the chimpanzee was trained to do it. Would it just waddle toward the flock, its long arms in the air, its lips curled back and shrieking, the sheep would simply stampede in fright.

Due to its lack of intellect, the chimp has a very short attention span and soon becomes disinterested. How often have we seen chimps in television set-ups dressed grotesquely in human clothes and being spoken to as though they understood every word? After blankly staring back for a while their attention span fails them and they often spoil the 'take' by shrieking and climb over the hapless human actor who has chosen to break one of the unwritten laws of show business by working with animals, and either gets urinated on, her (or his) wig pulled off, or suffers some other indignity or embarrassment, until the chimp is brought under control of the keeper. It makes one wonder when it will be realised that all these attempts to condition and humanise them will be deemed a complete waste of time.

Advertisements using chimps on television are the result of them putting together of a lot of patient and numerous 'takes', facial movements and expressions and film shots to make them appear to be speaking. Whereas a chimp will blankly stare at you in a rather 'wooden' fashion, the response from a dog is markedly different. An air of attentiveness, intelligence and understanding is obviously displayed. Dogs show an amazing capability to detect human moods and even their intentions. I have said before that anyone who has ever owned a dog knows the difference between your intention of getting up to make a cup of tea and the intention of taking him for a walk. The former will only raise a canine eyebrow, but the latter will have him jumping about well before you reach for the lead.

The high intelligence in other creatures, though recognised by anthropologists, is quietly ignored because their physiology and basic body form precluded them from any reference for 'cousins' or 'relatives'. When hearing parrots and mynah birds speaking and forming amusing sentences so articulately, it would amuse anthropologists greatly if, one suggested *they* were our relatives or ancestors, yet this is precisely what a certain Professor Huimar Von Ditfurth suggested in his book, when he stated that "Mankind and the 'chicken' had a common ancestor 'only' two hundred and eighty million years ago". [38] Clearly, he saw a common link with bird-like creatures and humans. Perhaps, instead of ape-like Pro-Consul common ancestor, we should be looking for a large bipedal bird-like creature, something akin to an ancient ostrich.

To return to the subject of high intelligence in dogs, a recent television programme called *One in a Million*, dealt with a canine pet that appeared to be sensitive to the onset of an epileptic fit in its owner and reacted accordingly by barking to warn her to become prepared for it. Sure enough, after the lady had taken suitable precautions the convulsions duly appeared. This is a very useful advantage to have for one so unfortunately afflicted, and can enable one to take precautions such as not being halfway up the stairs or standing on a chair dusting or handling boiling water, and so forth.

Although dogs do respond excellently to training, their extra sensory abilities seem to be in place as an additional advantage *before* any training, such as they become a guide dog, a 'tracker' or a 'sniffer', working for the law enforcement agencies, or even being apparently sensitive to paranormal phenomena.

A question many people have probably considered is could anthropologists have among their ranks ardently religious people? With all the afore-mentioned controversy and contention existing in their particular field of interest regarding human origins, they all at least unite and agree on one topic and that is the theory itself that humans had a 'natural' evolution from the primates. They will slavishly cling to this theory at all costs.

It would not seem possible for a paleoanthropologist who believes whole heartedly in the Darwinian theory, to be able to accept anything written in the biblical account of Genesis to have any likelihood of being possible. If some of them do feel they are quite religious, how then do they reconcile their beliefs? By taking out certain parts of the Christian religious teachings regarding divine creation and ignoring them, forming their own beliefs, believing this, but not believing that, rather like rewriting the Bible to so suit themselves?

Science is entitled to the view that the Genesis account for human creation defies scientific logic. Some might say that the biggest miracle that occurred in the Bible is nothing to do with those dealt with in the New Testament performed by Jesus, but the Old Testament account of creation itself.

But, having said all that, there does seem to be a certain common denominator in the strange situation that one could say appears to be like a mass racial memory, shared among all nations and their legends in that men were 'made' or 'created' by God, or the 'gods', as though some profound creative event did occur in our past and was passed down through all the earthly legends and seems to be installed deep in the human memory cells.

The simple proposition that all life on Earth may be the result of a freak biological accident and unique, even throughout the universe, possibly never to be repeated, deserves some attention and has many followers, but they would have to ignore so may logical conclusions that it would be unlikely but it would not be a barrier to accepting Darwinian theories as human appearance and ancestry would only be a natural follow from the initial process, i.e. evolution from a common source (proposed, incidentally by an ancient Greek, Anaximander, two thousand years before Darwin). In any case, when we are told that there are more stars in the universe than there are grains of sand on every beach on Earth, then multiply that by eight or nine for the planets, other world beings must exist.

However, the 'biological accident' so utterly defies the laws of probability regarding the likely existence of other similar planets to Earth, that it can confidently be put aside, as we know for a fact that, as many tons of cosmic material are deposited on Earth every year, there is no good reason to suppose that it does not happen on other worlds and therefore 'life particles' that are said to exist in meteorites may be regularly deposited there also.

The end result of any theory pertaining to human origins is the fantastically developed and over-endowed human brain. Even a pathological idiot, who can dress himself, feed himself and smile when he sees something funny on the television, or looks at the Sun or the Moon and wonders about them, is light years ahead of any chimpanzee.

We take so much for granted that we see all around us, but everything we see is designed, either built, created, planned for, cultivated or grown by humans. The brains that probed into and solved the workings of the universe, of gravity and its laws and peered into the smallest particles of matter, working out complicated mathematical theories and equations, can by no stretch of the imagination be seen as having anything to do with apes.

People possessing, in some cases, super-intellect, can only be described as massively over-endowed, and defiant of any evolutionary and natural selection laws that simply do not allow for such things to happen. With the *additional* brain material still to come into use, something or someone has provided a fantastic legacy to be drawn upon when the time is right, and that is something natural evolution would *never* do for *any* earthly creature.

It is easier, therefore, to view the amazing feats of human intellect more as a gift from God, or even the 'gods', than anything bequeathed from ape-like ancestors.

In terms of creeping evolutionary change, the recent development of the human brain should have taken millions, perhaps billions, of years to have occurred rather than thousands, its major development and achievements have only occurred with the development from Cro-Magnon to modern humans, after the onset of human cultural evolution.

It is noticeable that there seems to be a strange reticence and reluctance to even enter into any discussion or dialogue, that might suggest that there are a few doubts creeping into the theory, although such doubts are reflected in written works, they do not seem to be discussed on an open forum. Those who are active in the field (assuming they are still searching for the vital fossil) hope that all will be resolved when something profound and conclusive is found. An event of course, for which there is no guarantee. If the positive links where discovered no one could dispute many would embrace the 'natural' evolution of man from the apes. Darwin gives us much more than that, so we cannot detract from Darwin's intelligence and great mind. We simply make the point, that one need not have the intellect of an Einstein, to simply reflect or even question the part of the great man's theory regarding the emergence of the human, which simply does not fit the pattern of the theory and even offer possible alternatives provided they are sound and scientifically possible.

Darwin's brilliant work in the field of botany and so-forth is unquestionable, also his natural selection of the slow cellular construction of a complicated plant by slowly adding and improving its cellular structure. Adding up all that is good and rejecting that which is bad, obviously reflects a 'force' or a 'designer' if you will, what is does, is to provide an alternative designer to the creator of the universe and everything in it that we call God. As said, one simply uses the name 'natural selection' as the unknown force than that of a divine creator but the alternative does not disprove the former, it is simply a preference or choice.

Try as we might, we cannot get away from the fact the 'explosion' of human intelligence and a refined body form defies natural selection and normal evolution and is more indicative of a kind of purposeful creation.

The strange thing about human evolutionary theory as it stands, is not so much that it *may* well be true and, perhaps the ancient Pro-Consul bones *may* eventually be found, but rather the conviction among the searches that it could not be anything else but true, and although the continual dissent and argument continues to exist among those active in the field, they all unite on this one factor. In other words, the theory is totally unassailable. It could almost be said that the entire theory is becoming a 'sacred cow' and the longer the necessary proof evades the searches, especially with the aforesaid reticence, almost fear, to question the anomalies within, the more likely it seems that it will become an unqualified 'dogma' and subscribed to by an equal amount of blind faith as the Genesis account.

No other theory has ever been subscribed to for so long and with such compulsion and commitment, by any branch of science previously with such paucity of evidence. No doubt this is due in some large measure of the dramatic impact it had on peoples' lives, when the full implications of it became apparent, and it would be unthinkable *now* to entertain any doubts or dissent regarding it with so many people over the last one hundred and fifty years having turned to it from previous beliefs.

When certain anthropologists frequently cite the alleged human-like qualities of the chimpanzees, their blood, DNA and so forth, they ought perhaps to remember the late medical anthropologist, Sir Arthur Keith, who, as pointed out previously, identified three hundred and twelve distinct physical traits that set humans utterly apart from their so-called cousins. Among these, of course, would be the simple human ability to smile and almost speak by small facial expressions, the almost spiritual feelings of self-awareness, and an unidentified 'destiny' with amazing enquiring creativity and intellect. Other creatures, of course, have their own distinct characteristics, but none comes anywhere near the human entity.

Charles Darwin was not at all comfortable with every aspect of the 'apes to men' theory, but perhaps more so than Alfred Russell Wallace, the co-developer of the theory. They were both honest enough to make these doubts clear and, regarding certain aspects, they would make remarks like …. "As to this problem, I have no immediate answer". However, the general inference was that they confidently expected time itself and their devoted followers to eventually provide all the answers. Whether they foresaw such a lengthy period of one hundred and sixty years would go by with still many of the answers not forthcoming, is rather doubtful.

Mr Wallace was fond of the term 'some unknown variable', and Charles Darwin used a huge amount of phrases in the subjunctive mood, such as 'let us assume', or 'we may well suppose', and so forth. But for all that, it was a very brave act, not only because they were well aware of the backlash and furor that would obviously occur, but because, as previously said, not a single ancient bone of an alleged distant ape-like ancestor had, at the time of publication, been found.

It is strange that, at both ends of the enormous time period of six hundred million years ago to the comparatively recent appearance of anatomically Homo Sapiens, there are questions and anomalies with the rest of the timescale producing its odd occurrences throughout the intervening millions of years. We struggle enormously to identify any creature that could be singled out as a distinct transitional stage from one species to another (as said) to have the lung fish that left the ancient seas being the progenitor of *all* forms, it is obviously necessary to obtain and identity skeletons and remains of such examples of change.

The Earth sciences have not explained the afore-mentioned Cambrian explosion of fully formed creatures, yet no 'transitional' stages, the full explanation of the mysterious Hominid, or the mystifying demise of the Neanderthals and rapid appearance of the Cro-Magnon people. Clearly, the whole process of the theory is replete with 'missing links' and unanswered questions.

The so-called 'Cambrian explosion' baffled Charles Darwin himself and he did admit to being puzzled about why we do not find rich fossiliferous deposits belonging to the periods prior to the Cambrian. He made it clear that he could give no satisfactory answer. How *could* he, in his time, when after nearly a century and a half of scientific advancement later, even *we* cannot provide the answer, and so we are stuck with it. How *do* we explain the fact that the first part of our evolutionary history seems to be missing, as at the onset of the Cambrian system the initial divisions of species of both the plant and animal kingdom had already established themselves on Earth?

If we reflect on the enormous amount of time since the early appearance of life forms, which led eventually to humans and apes, then look at the amount of brain material that is 'neurons' or brain cells, that the ape possess in relation to humans, with their additional billions of brain cells, it can be seen that humans should have required an enormous amount of extra evolutionary time for the brain to have developed.

As said, possessing ten times as many brain cells as apes, humans would require ten times their length of evolutionary accruement. It is fact or such as these that encourage the development of alternative theories, even those bordering on the realm of science fiction.

Could the human brain be possessed of 'donated' genetic material that had an evolution of its own from Earth? We will deal with this in due course.

Brain sizes, and the various weights of the brains of certain creatures such as elephants and whales, seen irrelevant. What seems important is the brain weight to body weight ratios. When asking previously whether we should consider a distant bird-like ancestor, it is interesting to note how similar is the ratio of brain weight in man, i.e. 1:30 to the *sparrow* i.e. 1:29, the chimp 1:75 and the whales 1:15000. Clearly, the brain – weight ratio of the sparrow is closer than that of our alleged cousins, the chimps. [49]

Alfred Russell Wallace was greatly troubled by the wonders of the human brain. His was a time of expansion of the noble human achievements and empirical expansion, of great engineering feats all around the globe. It was the age of scientific flowering and advancement, with man stamping his dominion firmly upon the world. He must have reflected often on the chimp's blank stare from its zoological enclosure and found it very hard even to contemplate any attempt at comparison of the human intellectual capabilities to that of our 'cousins'.

Charles Darwin was no doubt equally as troubled, but not, it would seem as greatly as his contemporary A R Wallace. In *Origin of Species* Charles Darwin said …. "Long before the reader has arrived at this point in my work, a crowd of difficulties will have occurred to him. Some of them are so serious that to this day I can hardly reflect on them without being in some degree, staggered".

Clearly, modern humans with this amazing over-endowed cellular organ in the skull, are nothing other than baffling misnomer, an anomaly, a misfit and quite disturbing to an otherwise logical and acceptable theory that one assumes, including the human 'mystery', will eventually become proven fact, providing the necessary fossil material *can* eventually be found.

It is not only the current brain development of the average human that has to be considered, but also the many human geniuses that have lived and died are still to come. Then there is the problem of the alleged additional and unused brain cellular material that we could assume may one day be utilised. What is its purpose otherwise? Why was it provided? By whom, and how? Evolution does not 'over-endow' the theory of evolution and natural selection simply does not allow for such over-endowment.

The theory is more attuned to a slow step-by-step gradual and natural increase in brain material, with each evolutionary approved entity building and improving on previous inadequacies, but the fossil evidence shows no such thing. However, due to the fact they only portions of skulls are found and painstakingly stuck together, the brain cubic capacity has to be mathematically calculated and glaring anomalies arise.

The most obvious one, and most easy to observe, is the strange anomaly of the Neanderthals having such a large brain (1,400cc – 1600cc) in comparison to their successors, the Aurignacion/ Cro-Magnon peoples, who had less brain material, but slightly larger brains than humans of today. As if to compound this mystery of erratic brain development, a few skull portions are found and dated as *two hundred and fifty thousand years old* and called 'Swanscombe Man', with a brain size equal to modern humans. Again, subject to the validity of dating processes and correct calculation of the brain size, the Homo Erectus entity only had some 900cc of brain material, so a jump of 600cc of brain material in indicated in the time period from Erectus to Neanderthal 'Man' totally defying the normal snail pace of evolutionary change and genetic stability, and clearly indicating that the human brain development is only measured in thousands of years, when such a supremely gifted and sophisticated organ should have taken millions of years, by evolutionary standards, to have developed.

Other anomalies exist to bedevil the theory such as, if the evolutionary processes had struggled for a couple of million years in producing the various Hominid entities (who did have somewhat larger brains than the true pongid apes), and finished with the Homo Erectus entity, why did the process allow this supposedly more advanced proto-human to expire and become completely extinct and yet favour the supposedly disadvantaged creatures (the apes) to flourish? This seems to indicate that 'Erectus' had no real advantages at all, and was most likely just another variety of primate.

Then we have the afore-mentioned assumptions by certain anthropologists, who had him, boat building, adventuring and even navigating the Pacific to Australia, yet leaving no bone fossil evidence of his existence there. The entire range of alleged ancestors prior to anatomically modern Aurignacion men, have all completely and utterly disappeared.

There seems to be no good reason why these creatures, at least in 'pockets' here and there, could not have survived in remote areas, much like the natives in the interiors of Australia and Brazil or, if they could navigate the Pacific, there are plenty of islands there where they could have existed in peace for thousands of years, totally undisturbed without any natural enemies. When the age of exploration began, there is no good reason why they could not have still been around for the early human settlers to have encountered, such as in post Columbus America, or by the Pacific navigators and explorer. There is a modern theory in existence that suggest that the mysterious, so-called 'big foot' creatures are such an example, whose habitat is continually encroached upon.

There seems to be some confusion within the various sciences of the theory whether any kind of 'guiding force' exists or not. Some will support the notion that the evolutionary process is entirely random and impersonal. Others will see what they term as a 'watchmaker' busily putting together all the best pieces. Charles Darwin used phrases that suggest a force of some kind is firmly in control. "Daily and hourly scrutinising and rejecting, which is bad and preserving and adding up all that is good". The mystery is our inability to identify or quantify it.

Clearly, anything that 'scrutinises', 'rejects', 'preserves' always 'working' etc., could hardly come under the description of a random or blind chance process.

Dr Robert Broom, a past renowned paleontologist, would hardly make any friends among his scientific contemporaries if he was alive today to make his statement "it is clear to me that evolution was accomplished, not by natural selection or mutations, but by 'spiritual beings' of various grades and various kinds of intelligence". [40] it is clear that he found a sort of compatibility with the theory and his spiritual beliefs.

Norman Macbeth said, "The vitalists and other persons who see a 'watchmaker' or the hand of God, behind the marvels of nature should not be reckoned fools. They *feel* this presence and the Darwinian arguments are not persuasive enough to overcome this feeling". [41]

Today, one could say that the theory of evolution is *itself* almost treated like a religion and we could agree, as previously said, that an equal amount of blind faith among its followers is also necessary at this point in time, to give credence to it until, of course, the final proof in convincing skeletal evidence appears on the fossil table.

There is a wide diversion of human types on Earth with regard to the various blends from the three primary classifications of Caucasian, Negroid and Oriental, but the skeletons in appearance, in spite of the profound differences in height and stature are mostly the same. However, this is certainly not true among the primates, and when we consider the obvious differences in the range of Hominid skulls, from the grotesque to the gracile, this seems to suggest they were little more than additional primate varieties.

Where then did this amazing brain in the human come from? Not one preceding entity could be seen as a possible donor. If apes cannot even count their fingers after forty to fifty million years of evolution, it is hardly likely that a Hominid would have any inclination or ability to do so. Human finger counting progressed to theories and mathematical equations even those expert in the field find hard to follow. It is said that Einstein could only converse, regarding his theories, on a one to one basis with only a handful of contemporaries. Mathematics and one's ability to apply it, applies to everything in the Universe.

Sixteen scientifically named geological periods have passed from the Cambrian up to the beginning of the Pleistocene period with no Hominid or Proto-Human fossils evident within them, when the easier simple calculation of ape/human brain development should have seen early human-like fossils in *all* of them. Even geological disturbances, such as earthquakes and floods, cannot explain it. If such an abundance of dinosaur bones can be found, they also would have been subjected to such forces.

Among all the aforesaid three hundred and twelve distinct human characteristics *the* most important and obvious one, therefore, is the awesome human brain, with its amazing capacity for mathematical equation, which is really a totally *unnecessary* attribute for mere survival, which after all, is all that impersonal evolution should have been concerned with, but *very necessary* if humans were ever to travel in space.

As one anthropologist put it, with regard to the unexplained (and strangely un-discussed) anomalies, regarding the human entity, "Man may have come down from the trees but certainly is not yet out of the woods". Alfred Russell Wallace saw the human brain as an instrument developed in advance of the needs of its possessor and postulated on what it was that must have been at work in the elaboration of its extremely rapid development.

Clearly then, the dynamic forces responsible for the lightning and profound brain development in the 'post Homo Erectus' human, with language, upright posture, manual dexterity and supreme creativity, put the unique human light years ahead of *any* creature on Earth, and it staggers the imagination to think, with our current speed of technological advancement, what humans will be capable of it surviving for the enormous period that our 'cousins' have *already* had, that is forty to fifty million years of further development.

Cultural evolution among humanity only really began some forty thousand years ago, yet compared to the enormous amount of time that has passed since the Cambrian period, this is like yesterday. Undoubtedly then, the mysterious human entity is one of the biggest stumbling blocks to the theory. Certain anthropologists have at least tried to address the problem, such as Loren C

Eisley with quotes such as, "Measured in geological terms the brain development appears to have been surprisingly suddenly, a huge mushroom of a brain, which has arisen magically between night and morning. When I said that the human brain 'exploded' (into being), I meant no less". [42] The secret of how and why, certainly fits comfortably within the theory of an 'outside force, or extra terrestrial genetic manipulation.

Of course, in spite of its amazing and positive achievements and capabilities, the brain has a darker, more negative side, that is responsible for more questionable human behaviour that, for the most part, is held in equilibrium, rather like the nucleus of an atom, but the brain is already suitably pre-equipped with the ability to cure itself and, sure enough, all the genes responsible for mental maladies will gradually be identified, isolated, controlled and manipulated. So too, will all the other problems that beset the human, be suitably controlled, such as genetic obesity, excessive height (or lack of it) and all human maladies and ailments. *The process has already begun.*

However, this wondrous organ could be our salvation or destruction, and has come, at certain periods, of recent history, very close to the latter. Nevertheless, it seems likely that the rational side will prevail and overcome the dark side of the human psyche and that science, for all its faults, will ensure all those spears eventually become plough shares. Moreover, in this regard we could assume that advanced alien beings may have once possessed these negative aspects themselves but have now mentally evolved.

With regard to these very obvious mental over-endowment and highly unique qualities, only apparent in one single species out of the millions that exist on Earth, it has been said that the existence of such a creature as the human is a mathematical improbability, or he should not exist at all, particularly in terms of evolutionary laws. Natural selection laws do not allow for 'favourites'. We always return to the same problem that disturbed A R Wallace and his 'unknown variable' to explain these issues.

Why then does man possess this amazing excess of neuronal cellular capacity, over and above his simple needs to survive up to and remain within a 'brute' existence? These attributes are totally unnecessary, as previously said, for simple every day survival pursuits. Man is certainly worthy of the term' evolutionary misfit' and the reason for these strange anomalies seem unlikely to be answered by science, at least in the near future. We will just continue to take them for granted and utilise them to full capacity, in pursuit of our advancement up to the point when some unused portion of the brain will stir itself or we receive enlightenment from the donor. If it was not evolution that picked out man from the vast array of other earthly species for these special endowments, what, or perhaps who, *was* responsible.

I said previously, that Neil Armstrong's historic comment when stepping out on the Moon in 1969 could equally have been "That's one small step for man and the *third* giant leap for mankind", simply because the first two leaps before this monumental event were entirely responsible for making it happen, and they are, of course, the discovery of the use of fire and the mining out and smelting of metallic ores.

Research into brain impulses, its rhythms, and the various experiments in the field of neuronal activity, seems to ably confirm that humans are the only creatures or species of mammal that can imagine, pre-plan and visualise or create imaginary scenarios, before actually creating them and all this is just further confirmation of the uniqueness of the human entity in comparison to the rest of Earth's species.

Of course, this could be disputed regarding some factors, bees with their symmetrical honeycomb structures, beavers with their dams, termites and ants that seem to have organisation, but many say they are acting on instinct and survival rather than thinking and planning ahead. However, certainly it cannot be disputed that obvious intelligence does exist in other earthly creatures mostly if not always relative to survival. A sheepdog does not rely on aspects of survival in its highly intelligent capability to control a flock of sheep, it is a separate capability.

It would seem that whatever hidden force or mysterious factor (which seems to preclude natural evolution that was responsible for human intelligence), the initial plan was for an even more advanced and intelligent entity and these genetic 'clues' pop up from time to time with the occasional appearances of human geniuses and prodigies and especially gifted individuals that are above the normal endowments of the brain. They can only be described as massively over-endowed. The process of isolating genes, is the key to the ultimate control of all human traits, positive or negative, and could be described as a super weapon, to be utilised for good or evil.

Obviously, on the positive side, elimination of negative, troublesome or undesirable traits will be seen as beneficial, but we must bear in mind that such control will also encompass the negative aggressive traits who commit heinous crimes, and their only excuse is that 'voices' told them to do it. This isolation of such genes could lead to the eventual ability to create whole armies, if necessary, of 'cybernaut' entities, equipped in their mental areas with such killer traits to do our bidding in times of war or defence, with no compassion or other human traits that, in their case, would be seen as 'flaws' of failings in such entities.

Such entities would be devoid of all feelings of emotion or compassion and act rather like 'cybernaut' terminators.

To quote A R Wallace once again, "Exceptional intellectual activity cannot, by any stretch of the imagination, have become part of man's mental makeup through the process of natural selection that never endows a species with any particular desirable characteristics. [13] Whereas, other creatures of the world including all the foliage, seem to dutifully follow the laws of established evolutionary science, humans do not appear to do so; if they did, they would only live like the primates in a gorillas colony.

If it is true that we only use around one third of our mental equipment, or brain power, and leave two thirds fallow, then it would seem that the possibility of enormous intellectual achievements has been 'pre-programmed' and made possible in all of us, and if full usage of it is ever achieved, then the possibility exists that in the future there will be no unanswered questions for us to ponder about or problems to deal with as we will have surely answered and dealt with them all. Nevertheless, it could be a little disturbing to contemplate where it will all end unless we see is as our clear duty to travel the cosmos and assist other suitable creatures to evolve just as we were assisted.

On the subject of genius, this isolation of an intelligent gene, quite recently achieved, could be quite profound and very surprising simply because up to this point it could not be nurtured or encouraged or controlled in the sense that we have been able to regarding other traits and advantages in animals and things that grow.

If people of extraordinary intellect married, they could not expect or guarantee a 'wonder child' among their offspring, as it *has* actually been tried. Surely this is just another example of the human not following the laws that seem to work quite well when attempting to produce increased or better yields in farm produce, or in the special breeding of dogs, or in the production of find race horses and strains of cattle, and so forth, this is really quite surprising, because it indicates intelligent genes do not comply with hereditary processes.

This would appear to suggest that exceptional intelligence is a gift and not controllable by man. Surely, the isolation of intelligent genes will lead to nothing else but full control and mastery in genetic experimentation with them, perhaps even introducing them to *other* creatures. There is a border line of ethical restraint that human geneticists approach from time to time and, due mainly to adverse publicity, seem to agree to go no further, and if they argue about it there is an uprising of public opinion which obviously would not arise in countries that chose not to publicise what they were doing in the first place, and in any case, unfortunately, ethical restraints do seem to be dissipating in other aspects of human behaviour and particularly in the entertainment industry.

However, there appear to be certain 'grey areas' where the laws of restraint seem either non-existent or vague. Furthermore, some countries differ in their views on the whole question of ethics, for example, the broad brush and rather permissive statement in certain Eastern doctrines of 'the end justifies to means', virtually says in other words, "You can do whatever you like as long as it achieves your goal". This pretty well sums up the doctrines of (the once) more totalitarian eastern bloc regime of the USSR.

Studies in the genetic cause of mental health problems and birth deformities and so forth, may inadvertently cause the researchers to stumble accidentally upon the means to achieve other things that were not part of the original plan or goal.

We cannot guarantee, without specific and agreed international laws, that every single country in the world will behave in an open and honest way by voluntarily observing a noble code of ethics until such laws are in place and, more importantly, monitored for observation. In the film industry, the censorship laws were happily disposed of and of course in theatrical state productions, only common decency prevents excess.

Although the study of genetic science and its relation to human behaviour is a comparatively new field, the aforesaid animal husbandry and breeding programmes have been going on for centuries in many different countries and the breeders have simply followed a common sense approach, that their activities would produce the kind of animals there were looking for without bothering or thinking about the physiological or hereditary effects on the genetic code, and so forth, but once these studies became more intensified and begin to be applied, then obviously some form of agreed international restraints, or controls, will be necessary or a horrifying futuristic scenario of all kinds of questionable activity could be envisaged, particularly in regard to the proven ability to 'clone' certain animals and there is no good reason to suppose that such operations will only stop with animals, and for all we know such practices may already have occurred in a country perhaps least expected. Therefore, how could we guarantee for sure that such furtive and dubious experimentation has not already been carried out? The same ability could be in place to breed a race of Adolph Hitler's as well as a race of Einstein's. Therefore it seems obvious to assume the furious rate of advancement in the field of genetic experimentation, particularly with its regard to human undesirable traits, will proceed along the lines of 'being necessary and proper', but will also quite possibly act as a 'cover' for other more questionable activity due to discoveries made along the way, when working in the field of eliminating those genetic maladies.

What does all this imply? Well, we might consider the science fiction like, but nevertheless possible theory, that humanity *itself* could have been the subject of some form of advanced genetic experimentation on early man to enhance intelligence and the end result was the mysteriously over-endowed human brain.

There is in fact, another postulation among some scientists who have considered the possibility that life on Earth may have been nothing more than a freak biological accident and the great diversity of life forms on Earth simply progressed from single lifeform where all those ingredients coincidentally came together. However, the laws of probability would seem to refute it when one considered all the life particles in the molecular groups detected in space and the host of sun-like stars being detected and planets seemingly positioned in the ideal position or habitability zone.

It must be said that the spontaneous assembly of all the vital amino acids to produce life forms, that is, multi-cellular re-producing life, eventually gaining and developing intelligence, remains a high order of improbability but the fact is the necessary molecular groups are there in space, and an inestimable number of planet exist for such a process to begin.

Our solar system has only produced one planet suitable for life; many solar systems may not have a single planet that is suitable.

To consider the 'Fermi Paradox', named after the astrophysicist who coined the phrase. He postulated when the calculations carried out during the SETI programme seem to indicate the existence of other intelligent creatures, on other world, no doubt much older than ours, may have reached the ability to explore space, the question asked was "Where are all these people"? we are progressing rapidly toward space exploration so why not them? On the other hand, space scientists such as the late Carl Sagan did state that certain factors seem to indicate that such creatures may have visited Earth at some point in our past.

Stars like our own are something of a rarity. The general pattern seems to favour binaries and triple system. We must consider the astronomical forces that surely, exist in such systems and their effects on any planet struggling to generate life. However, on the plus side if any flora had managed to flourish, all that extra sunlight would certainly favour the photosynthesis process in binaries or triple systems.

We must also consider the 'Panspermia' process where amino acids reach Earth from space in meteorites. Amino acids are the building blocks of living protein although some have, it is said, a molecular structure different from earthly types.

Nevertheless, two scientists of the Arizona State University independently examined a meteorite that fell near Murray, Kentucky, and detected the presence of eighteen of the known amino acids. They also found the pyramidenes that are basic ingredients of the nucleic acid vital to living cells.

Now that we are in the new millennium we, also approach a time period of one hundred and sixty years since the 'apes to men' theory was first initiated. When we consider the amount of archaeological digs going on all over the world as well as the purposeful excavations and searching for our elusive fossil links, we are tempted to speculate if the necessary bones are there to be found. We can postulate on various reasons for the lack of finds. Such as geological disturbance, continental drift and once occupied habitats deep below the water by these factors would have affected other life forms that shared the planet with our ancestors and even further back in time to the dinosaurs. Yet we have no trouble in finding their ancient remains.

If we look at slow, plodding evolution, current humanity seems the best it can do and, with regard to comparisons with the basic apes and the higher human qualities, it *has* achieved quite a lot.

Will natural evolution of the brain cells finally result in a new enlightening? Will we suppress all our primitive urges with the help of a great expansion of technical knowledge, due to the flowering of the additional neurons in the brain, greatly assisted by advances in genetic science? The super sequence machines will identify all the genes for intelligence and all those responsible for negative, questionable behaviour and eliminate them. The flowering of a new and enlightened age could begin.

Chapter II

THE REAL ANCESTORS

Where I have mentioned the 'Cro-Magnon' peoples in this work, I mean the anatomically modern human remains dated around thirty thousand years old, found at the rock shelter of 'Cro-Magnon Les Eyzies Tayac-Siruil Dordogne' France. These beings where of the Magdalenian period or Paleolithic age. Of course, modern human remains have been found much further back in time.

An interesting article regarding these beings was contained in Bjorn Kurten's book *Not From The Apes*. He says, "It has been possible to demonstrate that human lineage can be traced far back where it still retains its unique character. The disconcerting fact is (to anthropology), that the basic human has always existed, not as the offspring of apes or primitive beings of the variety but as a man. Indeed, we may doubt that our ancestors where ever, what could be properly called an ape, and this makes excellent sense zoologically. The contrast between apes and humans in anatomy, are too great to be reconciled with a common origin and the same is true of behaviour".

Bjorn Kurten suggests that Neanderthal was a degeneration from the human character by severing their contacts with the ancient civilisation centers, then found themselves cut off by the ice sheets and glaciers that blanketed Northern and Central Europe and with this isolation and their limited numbers, considerable 'in-breeding' would have occurred and with a limited gene pool, the appearance of bad genetic traits significantly increased, leading to birth defects and physical mutations, producing their structural characteristics. Signs of severe osteoarthritis have also been displayed in Neanderthal skeletal remains. What this means of course is that humans have no ancient ancestors of the primate variety on a direct line of descent to humankind.

The Neanderthal creatures were said to have existed for seventy five thousand years, which, in evolutionary terms is extremely short, particularly when we find fossil evidence of the counterparts of other earthly creatures going back into the mists of time. For example the Coelacanth, a marine creature caught in a fishing net, fairly recently and thought to have died out four hundred million years ago, or the fossil remains of an octopus dated one hundred and twenty five million years ago, or a frog of one hundred and seventy million years ago. In *The Book of Life* (E Bury Hutchinson) Stephen Jay Gould says, "The Cro-Magnon peoples where contemporaries of the Neanderthals for a time and not their descendants".

Although Bjorn Kurten blamed the Ice Age for the degenerative characteristics of Neanderthal such as his squat muscular frame and beetle browed skull and bone deformities, his type of being occupied the Mediterranean and other areas that were largely unaffected by the ice sheets.

In modern books, journals and data regarding Neanderthal, it is hard to find anything positive that is written regarding them and must seem to refute the idea that he was a predecessor of anatomically modern humans; he was in fact an evolutionary 'dead end'.

The Cambridge University archaeologist Paul Mellors stated, "Nothing resembling Neanderthal DNA has ever shown up in humans". They are no more complimentary, regarding Neanderthals predecessors, the entity called 'homo erectus'. The paleontologist, Alan Walker stated, "There was no human consciousness in homo erectus. He was not one of us and did not think like us".

Humans could claim that they have no ancestors at least, of the primate variety. The Cro-Magnon was just like modern humans in appearance and had all the pre-requisites for further development and advancement. He was contemporaneous with Neanderthal who suddenly degenerated and disappeared.

An article from the magazine *Science Focus* was printed in the Metro Newspaper on October 3rd 2019. It said, "Around forty five thousand years ago Neanderthal and modern humans co-existed in mainland Europe, but over the course of the next five thousand years the human population increased dramatically enabling them to occupy new territories, while Neanderthal gradually died out. Exactly why this was the case, has long eluded archaeologists. Now an international team of researchers may have found the answer, modern humans developed projectile weapons, such as spears, bows and arrows, the enable them to hunt more successfully than the Neanderthals. As a result, modern humans had readily access to food and where able to thrive, whereas Neanderthals struggled to find food and eventually went extinct".

The team-uncovered evidence of projectile weapons dating back to forty-five thousand years (more than twenty thousand years earlier than any found previously) in the Grotta Del Cavallo, a limestone cave, found in Southern Italy that is known to have been used by early modern humans in the Paleolithic era. It was rather like a workshop. The article serves to show what many of us already knew, Cro-Magnon was highly intelligent, imaginative and superbly creative entirely due to his extremely advanced brain. The weapons mentioned in the article would not have been invented overnight, but through gradual development and advantages could be worked out and implemented. Cro-Magnon did not depend on his normal muscle power to deliver the spear, he developed an implement in the shape of a wooden shaft where the throwing end of the spear was inserted into a notch that virtually gave him an extra 4 feet or so to his arm and when launching the spear it was as though he had a 6 to 7 foot arm giving much force and distance to his throw. It is clear that Cro-Magnon understood the advantages of leverage and utilised such advantages.

The author Andrew Tomas in his book *We are not the First* studied the history of Cro-Magnon and stated, "Our immediate progenitor was 6 foot, intelligent and good looking and had a continuous existence until the dawn of history. He became the fore father of modern man".

Of the Neanderthal, he says, "He was only 1.6 meters tall, with short muscular limbs, broad chest and small forehead, with almost no chin and in comparison with Cro-Magnon Neanderthal was quite ugly". Cro-Magnon was, as said, contemporaneous with Neanderthal, who he removed from the European scene by his refined attributes Neanderthal was not a relation of Cro-Magnon, the Paleolithic paragon of intelligence.

With regard to Cro-Magnon, it is the speed of development and indeed his appearance in the first place, that makes him such an unexplainable enigma in terms of the evolution and natural selection model. The belief by many, that Cro-Magnon was a purposely-created being with a high infusion of intelligent genes, may not be far from the truth and we all carry his genes.

If a so far unidentified higher intelligence did create our Paleolithic wonder men, that same intelligence could also have genetically manipulated Neanderthal to his rapid demise. However, for all that, it could be said, that a certain 'degeneration' appears to have affected the Cro-Magnon during his lengthy development to modern day humans and that this reflected in the base, negative, and criminally evil people at large in the world today. This would have began with tribal warfare, utilising their weapons originally designed for hunting, then in individual killings, eventually leading to whole armies and different lands, partaking in war, that continues for centuries. The situation with regard to pollution began with the industrial revolution and was exacerbated by fuel driven engines, particularly the early motor cars with high lead content fuel. Lead in high doses can be lethal. Drinking vessels widely used by the Romans (historians have suggested) could have caused the insanity evident in some of the Roman hierarchy, such as Nero, Claudius and Caligula. Clearly, our earlier anatomically modern predecessors did not kill off the Neanderthals, although they had the means to do so. They behaved in a way more fitting of intelligent humans and appeared quite relaxed with leisure time, in order to develop their pigments and artistic talents, so evident in their magnificent cave art. They were not pre-occupied with survival from day to day, and lived in peace with their contemporaries the Neanderthals, which, as said, they could easily have eliminated. Exactly when the negative and evil tendencies and obviously the particular genes affecting such later behaviour came apparent is hard to identify.

For decades, anthropologists have viewed man from the Magdalenian period as the archetypal 'caveman' unkempt, wearing animal skins and toting a club, however, all that suddenly changed. In a cave near Lussac-les-châteaux in 1937, two people named Leon Pericard and Stephanie Lwoff uncovered a number of flat stones dating from the Magdalenian period that showed both male and female individuals in casual poses, they wore robes, belts, boots, coats and hats. One drawing was of a young lady in a kind of trouser suit, on her lap was a square flat item, with a flap much like a modern purse. She wore a short sleeve jacket and neat shoes. She wore a hat designed to flop down one side of her head. Other drawings showed men equally well attired. Trimmed hair and beards well cut trousers and coats with broad belts and clasps. These finds were a shock and an embarrassment to anthropology and as such they quickly attempted to declare them as fraudulent, but their hasty judgment was overturned when they were authenticated later. The fact that they are an embarrassment is shown in the way they are stored away and cannot be seen, except by special permission for those with 'the right credentials'. It was felt that they may be too disturbing for public scrutiny. To highlight the fact that these anomalies of 'sophistication in the stone age' are not unique, prehistoric cave paintings from the Kalahari Desert dated from a similar age show light skinned men with blond beards and well styled hair with boots, trousers and multi coloured shirts, coats and gloves. Their clothing was probably of a very soft and pliable nature, buckskin or doe skin and so forth. Therefore, it would seem, that far from being 'cave dwellers', they used them as convenient shelters to complete their artwork and design and cut their clothing, probably eat their meals our of the weather elements. Sculptured statuettes and clever carvings on antler bone, that some, being so minute, that they suggested either magnificent eyesight in the 'carvers' or that they used optical aids. They also worked with ivory to make badges, clasps and other decorating adornments; no doubt, for male and female clothing and it is clear they were very particular about their appearance. There are constant attempts to make the 'model fit' regarding human lineage. It is well known that historians in particular, once establishing a pattern of thought and write accordingly, do not like to have their sacrosanct theories and assumptions disturbed. However, all of the above does not suggest any smooth transition from Neanderthal or inbreeding, and in any case, a male, as described above, would be extremely unlikely to be attracted to a Neanderthal female, particularly when so many attractive women existed among his own kind. Neanderthal was an intermediary species, but in any case, the contrast between apes and men in anatomy are too far apart to be in any way related or have a common origin and this is also true of behaviour patterns. An intensive study of the Neanderthals resulting in a 400-page book, ended with the comment, "They remain the subject of debate. Clearly one cannot be too pedantic about our human lineage at present".

This common assumption that the people of the upper Paleolithic Magdalenian periods all lived in caves is belied by the facts. In Le Grand-Pressigny in France, very extensive deposits of stone tools where found, many thousands in fact, scattered at various depths over 10,000 acres, yet there is not a cave in sight.

At a place near Charroux, in the hillside along the Charente River, there are upwards of fifty caves, yet intensive excavations have revealed no signs that anywhere inhabited by men, but they were certainly 'utilised' particularly for their very professional artworks.

The people of these times where clearly a relaxed and creative people and, as previously mentioned, also well attired and where not pre-occupied with survival from morning till night and has leisure time for higher pursuits and activity.

Their dwellings must have equally reflected their status as formerly described. Interestingly, in the Lascaux caverns, renowned for displaying the aforementioned artwork, there are holes in the rock that appear to have been made to support wooden cross beams that would have been part of a scaffold system that allowed the Magdalenian craftsmen to carry out their work. this system was still being used on a greater scale thousands of years later by the artists of those times (Michael Angelo for example) to paint chapel roofs. One hopes that their work will last as long as that of the Magdalenian artists.

The aforementioned cavern scaffolding, is ten to twelve feet above the floor. With regard to this scaffolding, the implication is clear, that such methods may well have been used in the process of constructing their dwellings. A certain professor Doru Todericiu of the University of Bucharest, stated "The history of architecture indicates that scaffolding did not precede knowledge of masonry and construction but rather the other way around and that the and that the Lascaux artists must have also had the ability to construct walls, and to deny this would be rather like saying that the candle was invented before humans learned to utilise fire". A lot of this enlightening data can be found in the interesting work by Rene Noorbergen titled *Secrets of the Lost Races* if still obtainable) published by 'New English Library/Times Mirror.

The Abbe Breuil and Professor Lantler in their books Les Hommes *De Lage De La Pierre Ancienne,* mention the finding of a prehistoric oven at Noailles, it was made of squared off stones, held in place by a kind of mortar. Kilns and fired clay, where also found, along with sculptured heads of bears and foxes, clearly, the people of the period had other skills along with their very proficient art where creating on a grand scale.

The appearance of those people of this period as described earlier, does not fit the pattern of the ideal evolution model, that is a gradual development from a simple being only intent on daily survival. The transition was far too abrupt, fitting more with the statement by one anthropologist "The steps through to the appearance of homo sapiens appears more like an 'experiment' than a desirable evolutionary hypothesis". One might also use the word 'creation'. A R Wallace 'Unknown Variable' leaves room for a form of purposeful genetic creation by advanced intelligence 'not of this earth' enhancing intelligence in a less endowed creature.

The Magdalenian folk exhibited largely, good health in their anatomy, with a noticeable absence of deformed and diseased bones as indicated in some Neanderthal finds, particularly the later ones. Perhaps, later when we look at the possible 'third alternative' the possibility of purposeful genetic manipulation, caused the Neanderthals to 'degenerate' as an anatomical failure, in order to clear the way for the development of the beings deemed as a successful operation especially in regard to the advanced human brain not only normally over-endowed but being equipped with even additional brain material, absolutely ruling out any evolutionary or natural selection process.

In Montgaudier, which is about 100 miles from the coast in France, a bone baton was found, that was engraved with depictions of two seals and a spouting sperm whale. They were so intricately carved that the seals can be recognised as male and female.

The implications of this are remarkable; travel was obviously not a problem for the artisans. Do whales come that close to the coast for close observation? If not, then it would infer sea travel capability, perhaps more enlightening revelations are in store for us as certain excavations continue. However, it seems that the aforementioned genetic manipulation was occurring worldwide. This is reinforced by the work of Professor J L Myers in the *Cambridge Ancient History* Vol. 1 page 48, where he noted conspicuous similarities between Aurignacion skulls found in Lagos Santa in Brazil and along the coast of Eastern South America.

Notations on a mammoth bone where marked off in graduations in a series divided by longer lines at specific intervals suggesting some special events at these points, or perhaps used as some kind of measuring rule, or maybe an early version of a 'calendar', indicating certain points of importance to the carvers.

Consider the following, Alexander Marshack, an American researcher, after analysis of the aforementioned notations stated "There is unmistakable evidence that it was a detailed record of Lunar phases, what is more, the notations pointed to its usage as a calculator, the phases of the Moon could be predicted in advance, indicating that Paleolithic man was a mathematician and an astronomical observer of his time.

The pre-historian Robert Silverberg states "Paleolithic art is upsetting to those who prefer to think of quaternary man as little more than a cave dweller. Not only do such discoveries indicate a great craftsmanship but they point to a whole constellation of conclusions: that 'primitive' man had an organised society, with continuity and shape. They had sealed heights of artistic achievement that would not be reached again until the late Christian era". We have stated that the ancient artwork of the Cro-Magnon, particularly in regard to dimension and perspective are breathtaking as also are the previously mentioned intricately carved artifacts that suggest they had to be completed with some kind of optical aid. Imagination and the ability to form clear mental pictures are a feature of Cro-Magnon's intelligence. He could hardly have done other than imagined the scenes relative to the hunting and grazing of the animals that provided the milk, food and clothing for his survival and then painting them from feats of memory within the caves. Nearly forty thousand years ago our Cro-Magnon artistic ancestors where ably demonstrating these gifts, with the aforesaid feats of memory in their detailed depiction of the animals around them at the time.

It is hardly likely that they coaxed or dragged the bison and woolly mammoths into the caves to pose for them. Another feature that portrays leisure time in their daily lives is the gradual and refined development of the various pigments and colours they used that are so evident and long lasting to be admired so many thousands of years after their completion.

There is no question that the Cro-Magnon peoples where a misnomer, an enigma, far removed from normal processes of evolution and natural selection. We have suggested that such an evolutionary advanced body form together with such mental attributes conforms more to a type of creation, but the religious factions would hardly accept two separate creations, firstly the big bang and the universe. Then billions of years later Cro-Magnon however, the Bible and genesis do depict humans that is, Adam and Eve, as anatomically modern there lies the dichotomy. Unexplainable in both ecclesiastical and evolutionary terms, thereby leaving the field open for a third alternative explanation which we have speculation within these pages. Clearly, the bible is quite specific about Adam and Eve being just like ourselves, but the writers had no conception of any relationships with former 'primate' beings with regard to the human. However, whether they realised it or not, they were saying that modern humans where created at some point up to one hundred thousand years ago. Yet the Bible teaches that all creation occurred at once.

Chapter III

GOD AND LESSER 'GODS'

The teaching of Jesus in the New Testament, even if it is disputed they may not be all His words, are wonderful, faultless and admirable. The biblical writers where quoting words of a very special person indeed. Why? Simply because they conveyed such profound truth in such terms that even the simplest person could recognise the analogies used in relation to their everyday life, and it is quite clear that they were a recipe for peace and harmony and they required a striving by all to implement the highest and most positive qualities in the human entity, and to suppress the negative (however, the new testament writers were a 'new breed', when comparing them to Abraham and Moses whose writings dealt with much mayhem, slaughter and purposeful disposal of masses of humans).

The obvious conclusion is that if everyone abided by them, then there simply would not be any problems in the world other than those caused by uncontrollable factors of geology and weather. There would be no murder, crime wars and slaughter, and man and his earthly resources would be free of such negative pastimes, to control the currently uncontrollable. However, it is quite clear, that when the New Testament writers where compiling their texts and gospels, most if not all of the commandments, had been ignored for centuries.

All negative activity on Earth is caused by human failing and the teacher (or teachers) of righteousness recognised the root cause of such failings and used the most simple terms to relate the most profound in their teachings. These admirable doctrines have been tainted by human folly and shortcomings and used as tools of destruction and murder in their name. the burning of 'heretics', the killing fields of the crusaders, self imposed zealots exterminating hapless old women on the evidence of children in the Middle Ages' witch hunt. Killing people in the name of religion. Many of the perpetrators where using the religious furor to implement their zeal in murder and bloodshed, almost like a return to the Old Testament activity.

These activities were a far cry from 'love thy neighbour', 'turn the other cheek', 'forgive thine enemies', and so forth, however, it can be seen that the Gospels and teachings of the New Testament are in direct contrast to certain events in the Old Testament.

The teachings of Jesus set down in the Bible, whoever may have uttered them, are very profound and instructive with regard to encouraging righteousness into humans in general. It was as though a code of behaviour was being gently instilled into the people and with many it most certainly seemed to work, with various conversations such as Saul, a onetime oppressor and tax collector, and with followers willing to go into foreign lands risking their own lives, rather like the future missionaries to come, by furthering the Gospels in sometimes quite hostile lands. They were clearly, safer in foreign lands not occupied by the Romans. Nevertheless many died in Roman arenas, rather than renouncing their faith, especially during Nero's crimes. In the book *Exploring the World of the Bible Land* (Thames & Hudson), says, "Nobody would dispute that the works contained in the Bible cover a vast span of time and that they were written by many different authors. Many authorities would also agree that even within a single biblical text there is evidence that many different hands were at work at different times". No doubt, also, many texts where completely removed or altered.

Early in the 16th century, the zealots of the Middle Ages consigned William Tyndale to the flames for his translation of the Bible. There were many other editions before the authorised version of 1611, and of course, other further revised versions were to follow. It has also been claimed that many alterations by various Popes down through the centuries have been made to please various kings and emperors. During the conversion of the Emperor Constantine many revisions, removals and no doubt additions took place.

The astronomer Giodorno Bruno stood little chance of escaping the flames with his revelations of Earth not being the centre of creation if men could be burned for simply producing an unacceptable edition of the Bible. To reflect on the ignorance of the dark ages in comparison to former, much more enlightened times, Bruno was stating things that had been postulated upon in pretty much the same way by certain learned Greeks two thousand years before his time. This zealous, indeed murder, in the name of religion was never part of religious teachings (in the Christian faith at any rate), and only served to further alienate science and harden their resolve to prove their theories, whether they offended the Church or not.

Through it all, the human brain has been developing and improving and from all that has been said, it is hardly likely that any of the creatures that arose and expired with their brief appearance on the world stage, especially the unrelated Neanderthals had anything to do with these amazing bequests apparent in the move positive and admirable traits of the modern human. It is hardly likely that the Neanderthals' struggling (and failing) to survive had either the time or the inclination to sit around the campfire pondering and questioning their beginnings. All the various legends, many probably handed down by word of mouth long before the invention of writing, concerning man and his origins or 'creation', are all a product of modern Homo Sapiens. With such intelligence evident in the Magdalenian peoples, they must have pondered and discussed their beginnings.

Most people, if not all, would be quite comfortable with the notion that a divine creator waived a majestic hand and within seven days the entire universe, the Earth, and its creatures, were all installed and happily functioning here on Earth. Most, if not all, theologians probably now reject the creative events described in Genesis did occur in that manner and certainly science totally rejects it. Nevertheless, the theme of a creative event by God, or the 'gods', runs through all earthly legends like some form of mass racial memory and seems to appertain only to this period of modern humanity since the onset of cultural evolution. When we mentioned whether the early humans pondered their beginnings, it may well have been they that instigated the worldwide belief in 'gods' creating men.

The *scientific* event concerning creation, however, took place fifteen to twenty billion years ago. For many people who assume that this amazing event *was* caused by the hand of God, it would be something of a dilemma to ponder why such a creative God would wait for ten to fifteen billion years before singling out a rather mediocre backwood area of an unspectacular galaxy to bestow just one planet with such a high diversity of life forms, unless, perhaps, He had done the same thing on many others who *we* have yet to meet, after all, our solar system only formed less than five billion years ago.

Science is happy with its enquiry and ultimate decision that the 'Big Bang' did occur, but what fantastic process 'allowed' it to happen? The strange factor is that the appearance of modern humans with all those afore-mentioned qualities appears to be a unique and separate act of creation within the major event that ultimately produced all the coincidental legends, one of which has found its way into the Old Testament. Most biblical scholars will admit that some biblical stories may be based on events and legends from one of the oldest civilisations known, that is the Sumerians. In addition, we have to realise that a large part of the early history and scientific achievements of humanity have been destroyed through ignorance and fear.

The archaeological evidence from various 'digs' around the world, such as in Turkey and India, have produced evidence of very ancient forms of civilisation indeed down through history, but the term 'civilisation' is usually applied to a period of 4,000 BC or 5,000 BC, regarding that of Sumer, also known by other names such as Sumaria, Babylonia, Mesopotamia, Mitanni, Elom, Edom, Eden, Akkad and Assyria, but it is fairly certain that the well presented and well dressed Magdalenian peoples we have mentioned saw themselves as 'civilised and that was forty-five thousand years ago.

The significant discovery of thousands of clay tablets and the various interpretations of them by different people make for very interesting reading. Interpretations, for example, in a five-volume work by Zachariah Sitchin states that beings called the Anunnaki that lived among the Sumerians, were the creators of Homo Sapiens by advanced genetic engineering processes. The various interpretations of these Sumerian tablets form the basis for subsequent myths, legends and even biblical stories, for example, creation, the flood, the Tower of Babel and even the 'babe in the basket' story regarding the origin of Moses himself, assigned to be (according to the British and Foreign Bible Society) the writer of Genesis, human creation and the flood story involving the biblical Noah. The Sumerian account, however, involved a person named as Xisuthrus.

There was also an ancient Indian 'Noah' called Vaivasvata warned by Vishnu to build s ship for the coming flood. Even a Chinese version exists. Although it is said that Moses, or even Abraham, have not been historically proven to exist, the fact is that if he *did* exist then it would appear that he drew heavily on these older account. We might mention that the tuition of Moses was undertaken by the most knowledgeable people in the world in his time that is the Egyptian scribes.

However, all the stories of human creation from whatever source, all come from an event that is said to have happened with the, (or perhaps the *beginning* of) the period of anatomically modern humans, when the physiological capabilities for human speech (and therefore handed down legends) were in place. The antediluvian people, which of course would include Noah (if he did exist) clearly had very advanced knowledge and created the out of place artifacts suggesting high technology we find today. Various Romans (Caligula) have claimed the title Divine. They seemed to sense a greatness, or power, within themselves that they could not quite define. Some claimed the title, although as leaders of great nations, more through mental aberration rather than possessing the qualities deserving of such a title. The Roman Emperor Nero, for example, who clearly viewed himself, and who encouraged others to see him, as a divinity. Other great leaders who did not seek the title seem, however, to be more deserving of it, such as Solomon, said to be the wisest of human leaders, and his father, David. To mention Moses again, he stated in his flood legend that Noah was high enough to 'walk with the lord'. Moses had many years of tuition, as he was an adult when banished from Egypt.

As previously said, some scholars and historians will say that the chief patriarchs of biblical fame, such as Abraham, Moses and even Jesus himself, have not been proven historically to exist, but as said in the beginning of this chapter, whoever the writer of those profound teachings was, seems to be quite deserving of the title 'Messiah', this, of course, is relevant to the New Testament but for the actions of the 'Lord' and his angels we must read the words of Moses in the Old Testament, which differ markedly from the New Testament.

It is known that the Old Testament is largely compiled from even older Jewish accounts and the New Testament is the Christian Bible, but the Bible itself did not establish itself into any coherent written form until around one thousand years after the destruction of the temple in Jerusalem by the Romans in 70 AD.

So Abraham, in spite of statements that he is not (historically) proven to have existed, appears in the legends of two different lands. Further, a historical character that interacted and that was even said to be the half brother of Moses, certainly existed and he, of course, was the Egyptian Pharaoh Ramesses II.

Although the Egyptians may have taken great pains to eliminate Moses from their own historic writings, they do little about other countries wishing to record his deeds and exploits, and Moses seems well entrenched, just as Abraham is, in Jewish lore, such as the Talmud. Abraham, or Abram, is the common patriarch of both the Jews and the Muslims, with the Jews tracing their ancestry back to the son of Abraham, called Isaac, and the Muslims to Ishmael, Abrahams other son.

The Sumerians believed that Eridu is where 'kingship' or political society first came 'down' to Earth, including the arts and other accepted features of civilisation. Eridu was located near the ancient land of Ur in Mesopotamia, and in 1949 a mound was excavated there dating back to 5,000 BC, and or course, it was the birthplace of Abraham.

It was here that the legends say the 'gods' came 'down' to inspire and enlighten their creations with the finer points of civilisation, such as the art of writing and the building of the first earthly city 'Uruk'. Possibly the earliest temple yet discovered was an excavation near Urfa in Turkey, dated from around 8,000 BC, where winged half-human figures have been found. Many postulations have been made as to what the ancients might have seen to assign them wings, as birds were the only possessors of such attributes, enabling them to move through the air. Are we to assume that the ancients saw humanoid figures that moved through the air and, as they knew of no other way they could do so, drew them, or made figures of them as 'gods' or as in the case of the Bible, 'Angels' and depicted them with wings?

In the Middle Ages, satanic forces were seen under every bed and around every corner, and the same zealousness and hysteria prevailed hundreds of years later in the Communist 'Witch Hunts' of 1950' America. In the sixteen hundreds there was a person called Mathew Hopkins the witch finder general. He was nothing more than a murdering psychopath masquerading as doing the work of God, and clearly had church approval.

There is a tantalising reference to a teacher of righteousness existing though not being named as Jesus, in the Dead Sea Scrolls. Many would see the need today for another teacher of righteousness to come among us with our continual wars and various areas of oppression, conflict and starvation of peoples in favour of buying weapons to wage useless conflict. But would he necessarily fare any better than the original? Perhaps his execution would be brought about once again, but this time by more subtle means, such as a form of political or mental execution rather than by physical means. Derision, scorn, branded as a false prophet or even a lunatic. To be sure evil in all its forms goes entirely unpunished in most cases today where self interest and corruption override any real condemnation of oppression in certain lands, but no impressive mystic Messiah travels the land with a number of followers preaching and working wonders which today would have to be highly impressive, with our knowledge of illusionists tricks.

There is a strange paradox in the afore-mentioned wrath of God severely implemented in the Old Testament in comparison to the gentle forgiving methods of the New Testament, almost as though a new hierarchy in Heaven had a complete review of the policy, with God, the Son taking a wholly different line to the more severe policies of God, the Father. However, when remembering the ultimate fate of God, the Son, how would He have justified and claimed success for His new and refreshing policies on returning to Heaven, when they resulted in His own termination, and not only of His right hand man Peter, but countless converts, dutifully going to their deaths in the arena to amuse pagan Romans? In addition, if all that was not enough, his prediction of his 'second coming', never came to pass. When speaking of it, before his ascension He stated "This generation shall not pass until all the things are fulfilled". Obviously, a major decision or postponement regarding a 'second coming' was made in 'Planet Heaven'.

But we must remember those who did go into foreign lands to preach, did obtain many favourable results and the spread of the Gospels is apparent today for all to see, and what greater test of faith could there be than people laying down their own lives rather than renouncing their beliefs? They were very brave men and most certainly had been highly impressed by the miracles and actions of their 'Messiah' to go to their deaths happily singing as they did in Pagan Rome.

And so, the bottom line is, whichever way they are viewed, and whoever the writers were that produced them, the simple, but profound teachings, attributed to a man called Jesus contain all the answers to all the problems that afflict the world today that are entirely due to the negative behaviour patterns contained within an otherwise wondrous organ, the human brain. These teachings cannot be seen as anything other than profound wisdom and the only means of saving humanity from itself. Therefore, humanity could accept its guidance from the New Testament where the teachings seem more oriented toward parables one can relate to and are a guide to bringing erring humanity back into the right path. However, for all that, theologians by and large seem to avoid the glaring and profound differences in the largely peaceful pronouncements in the New Testament compared to the Old Testament.

Certain archaeological discoveries have proven the Bible generally to be a reliable data source, but the Old Testament is viewed as a compilation of stories and major dramatic events from human creation itself to the flood, and the destruction by the wrath of God of iniquitous cities, the exodus and lives of the various kings, and so forth. Its very age and long history of variation, intentional or otherwise, may contain many distortions of original events. Theologians might ponder on the anomaly that an infallible almighty God created beings that later were seen fit only to be destroyed. In other words, he created 'mistakes'. Even earlier, destruction of the *entire* human stock seemed necessary in the story of the flood, except, of course, for a chosen group. This seems to imply that mistakes *were* inherent in the creation plan, unforeseen at the time, but necessary to correct later, which of course, was not the last of destructive eliminations of undesirable humans.

A book on biblical events says "Genesis treats the creation of life as an 'experiment' on God's part". The question must therefore be would a divine infallible God need to 'experiment'? Those who might subscribe to the 'alien biogenic creation' theory would also point to a kind of experimentation evident in the fossil record, with different proto-human types appearing here and there with various unrelated entities appearing and disappearing until the process seemed to settle for the Aurignacion, our Cro-Magnon predecessors.

Interestingly, as in past epochs, there is no other more refined version of a human genus 'waiting in the wings', so to speak, and modern humanity *has now been around as long as the doomed Neanderthalensis Sapiens.* Furthermore, when we look at the constant bloodshed, bigotry, inter-racial hatred, hostility and was manifested in the darker traits of an otherwise fantastic creation, we could assume that humans appear to be *still* offering themselves as candidates for destruction as a general failure. But we must say, that a modern day Abraham would find more acceptably behaving people to bargain with the ET's who may be considering another human elimination event, as in Sodom & Gomorrah he could not find any.

Nevertheless, as long as these traits are held in equilibrium there is the distinct possibility that the necessary scientific advancements will occur to eventually eliminate them, and so it is quite clear that the positive traits have a clear advantage in being able to dominate in the long run, after all we have been given enough extra brain material to accomplish almost anything when it (or if) it develops fully, and all it takes is time.

However, the negative forces will not capitulate so easily. When genetic science reaches the point of isolating the offending genes, the retaliation beings. People of influence will rise up and oppose their elimination. They will put forward valid arguments that suggest to eliminate 'carte balance' all such traits from the human psyche will reduce Earth's peoples to a mass of weak, saintly individuals who would be helpless in defence of their world to possible covetous alien aggression from space for example. But we have to say that if the events in the Old Testament really happened, then alien aggression was clearly evident in the past, as biblical writings confirm, particularly in the case of the slaughter of the occupants of iniquitous cities, such as Sodom and Gomorrah and the mass drowning and floods mentioned in the story of Noah and his alleged righteous group, seen as the only humans fit to be saved from the 'wrath of God'. Today, science is seen as more of a challenge to the Old Testament than to the New Testament, more specifically, the story of human creation itself described in Genesis. With all the alterations and removal of certain texts in the Bible, said to have taken place, the 7 day event for all creation should have been the first item to go, if those who did all the modifications where possessed with common sense.

The Church generally, has had a fear of science ever since the days of the alchemists and many would say that these fears were amply justified in the light of modern day advancement, where even the creation of life itself will soon come under human dominion. No 'creator' of humanity, whether divine or otherwise, would fear human advancement as both would be responsible for it. Therefore, it is clear that, (as harsh as it may sound) a divine God, the Creator, must bear some responsibility *Himself* for any challenge by man and his futuristic exploits in his ability to possibly create life itself, simply because such ability was given, in accordance with the Old Testament teachings by God, when he provided the creative intellect of the human. Many earthly legends are of similar content in many widely dispersed areas. Creation of humanity has diversified through the various legends totally agree that humans where created one way or the other.

It seems that the most suitable word we have chosen to differentiate man from the animals is his 'spirit'. That word separates him from other creatures of the world by an insurmountable and enormous gulf. Humans have an awareness, not only of themselves and their surroundings, with the wonders of nature, but of the entire vast universe, and possess amazing and irrepressible creativity, art, intellect and abstract though processes that no other creature of Earth possesses, and there seems to be no good reason why humans, reduced simply by some anthropologists to 'naked apes' should possess them either. *But they do.* How, then, did they come by them?

How then, do we solve the strange mystery of human origins? It seems preposterous to suppose that the human entity at its best could be the result of a freak accident along with all the other life forms of Earth. But it may be just so. We have no way of proving it without examining every other planet that may exist. Science still cannot prove conclusively that humanity evolved from the apes and they have tried for one hundred and fifty years. Theologians fare little better and cannot be sure at all that *all* the Gospels are genuine, or exactly as they were written long ago, and there have been many editions of the Bible.

Therefore, if we take the freak accident hypothesis out of the picture, what are we left with? Some people might take the view that after the one hundred and fifty year search for the proof with no results we may soon have to abandon the apes to men theory altogether and go along with Charles Darwin, who after all, did make the previously mentioned statement, "If the fossil links are not found then the theory falls down", and so we might eventually have to consider it out of the question that apes bequeathed the fantastic 'super brain' to humans when they have so little themselves, and *still* swing in the trees after nearly fifty million years of evolution from the earliest primates and have never created anything. When that time comes, we might have to just consider the final choices, that humankind were created by God... or was it perhaps 'the gods'? Why does the Genesis account instruct the newly created beings Adam and Eve to "Subdue and replenish the Earth"? If a Divine God as is taught, created everything else along with humans including the Earth, then it should have been in mint condition and not have a need to be 'subdued' but 'maintained' in mint condition, grow, reap, receive and put back to grow again, a need for 'subduing' would be evident to any visiting ET intelligence arriving on a 'planet of the apes' as it once was, the apes would have done very little in the way of subduing but it would have been a perfect state to assist them in escaping from predators.

Logically thinking individuals might suggest that surely an infallible almighty, all-knowing creator would be fully aware of all those who would make it and all those who would fail, right from the beginning, once again bringing forth the question, why create them in the first place? But, alleged ET 'creators', who are not divine or infallible, would not be exactly sure how their creative activities would manifest in the way they expected, but it seems to have been discontinued after Cro-Magnon.

There is also another anomaly that arises, if one chooses to ignore the more comfortable situation quietly accepting all the teachings without question (in short, blind faith), and that is the offering that God, *is* always *was* and always *will be*. This is quite a profound statement and it hard to accept, or define, when we try to analyse it in logical terms.

It is children that, for the most part and not just in regard to religion , that ask the most embarrassing, the most awkward and the most hard to answer questions on any topic. The above will usually be offered if a child asks where did God come from. And it is not surprising that the parent probably could not answer, but stating, "He is, was and always will be", is not a logical concept to many.

When pondering the suggestion that God *always* was, along comes science to tell us that the universe had a beginning and therefore time had a beginning and that this 'beginning' was some fifteen to twenty billion years ago, with the great explosion that began 'time', creation and existence.

That would eliminate any need to explain the embarrassing question, 'what was God doing in the vast amount of time (or eternity in reverse) *before* the 'big bang'?' According to scientific thinking, God himself should have been created *with* the 'big bang'. In other words 'God' is the universe, but other clever postulations come along regarding the possible existence of 'multi-verses', in which case a hierarchy of 'gods' would exist and now we are back in the Greco-Roman era.

Others might ask, how *can* time have a beginning, or be said *not* to have existed before the 'big bang'? Imagine the seething mass of hydrogen just a few minutes before the 'big bang', clearly, it was only a *matter of time* before it exploded, logically, therefore, time *did* exist.

Furthermore, those who now have an image of God sitting on his hands waiting for an enormously long time for fifteen billion BC (or before creation) to come along, may have the problems solved for them by science itself, who may now have answered this very question. The answer may be in the missing dark matter of the universe, which may have enough gravitational mass, along with the rest of the universal matter, to arrest the explosive flight of the galaxies and bring them all to a standstill, then have them start to gradually return to the starting point once again for yet another 'big bang'. This has actually been suggested by science, but one could easily imagine such a process creating such an enormous black hole that all the material of the universe would simply disappear down the gaping black maw, never to emerge, but the factor that opposes this is that the universe is increasing in speed rather than slowing down.

Nevertheless, the same conditions existed for a black hole before the initial 'big bang', and it would appear to have been possible for the event to have occurred then without any problems.

So science may answer the question as to what God was doing all that time, and that could be a continual overseeing and supervising of a whole series of 'big bangs' going on for the time immemorial. People like ourselves may have existed on similar worlds one hundred billion years ago.

There is no doubt that this mysterious 'big bang' is the point where science and religion meet. Was it a (perhaps continuous) divine act, or was it a scientific process that 'magically' produced all the hydrogen, seemingly from nothing? However, modern analysis suggests that as said the dark matter is 'accelerating' universal expansion. It was Moses, who (according to the British and Foreign Bible Society) wrote the words in Genesis regarding the religious creation story, and we have been puzzling over it ever since, but not those people who existed in Moses' days who revered him. Yet surprisingly, many people still, after so much advancement in our knowledge, accept it and are not at all concerned about their blind faith.

It is necessary here to stop and consider this amazing, but nevertheless possible, 'biogenetic creation' by an off-Earth creative Intelligence. At our current rate of advancement (minus, of course, the ability to travel interstellar distances), it will not be too long before humans themselves may be capable of such acts, given that 'Adam and Eve' appeared anatomically modern and thereby, as said, suggesting fairly recent creation, would an almighty creator wait for so long after his main creative work (possibly the 'big bang') to place humanity on Earth, and what was his reason for choosing that particular time period? Well, we might find it difficult to answer the second question, but the first question's answer might be because God is divine – he is possessed with divine patience, but this kind of explanation is not really an explanation at all, but just a way of avoiding trying to find one.

We could, of course, ask many other questions such as, how many other life forms have been created on other worlds. Will we ever meet them? Were *they* created with equal proclivity for negative and positive behaviour also? Which heaven will they go to if they manage to escape going to the other place)? How many heavens are there? Do all those other worlds have their equivalent of Adam and Eve? Were they in God's likeness also? Are the other life forms on these worlds also similar to Earth's creatures? Surely, we cannot all be the same throughout the universe. Is Earth *the* only world out of this enormous multitude, with life in abundance? The original estimation of there being one hundred billion stars in our galaxy has now quadrupled and we hear a figure of four hundred billion being stated.

Why has a divine God, capable of anything, created such a disorderly universe as we asked earlier? There is absolutely no point in such an operation, a construction engineer would not purposely design and create a high-powered engine with purposely built-in faults. These cosmic events in space are quite a serious nature and if there are as many planets that should contain intelligent life, they are bound to be affected by them.

Stars exploding and wiping out any planetary life forms, galaxies colliding, countless possible civilisations living on borrowed time, suns expanding gas clouds, etc. Closer to home, we have comets, meteorites and asteroids to content with and the threat of absolute extinction. Will the divine creator waive a similar majestic hand, as used in his initial creation, to vaporise the approaching mountain sized rock hurtling towards us? He would obviously have the power to do so, but has humanity earned their right to survive?

To return to the Garden. As Adam and Eve emerge, what racial characteristics do they posses? They have the formidable task of creating the three human varieties, Negroid, Asian and Caucasian. If subscribing to the alien biogenetic hypothesis an obvious question is, how did, or why did, the hypothetical alien creators produce the three distinct racial types? Did they bring the genetic material of these distinct types of human-like entity and then experiment with three types of primates – perhaps gorillas, chimpanzees and orang-utans? With all this sensitivity about race today and severe political correctness, it is surprising that there is no outcry regarding the white Adam and Eve.

Clearly, there are too many questions and few answers. If humanity in the distant future do become creators themselves, then it *would* be necessary to carry out certain tests on the created beings for their required characteristics. For example, their will power or their curiosity, which is obviously a feature of intelligence. Could this whole temptation scenario in the Garden be a distorted memory of such an event? The 'serpent' entity may well have been the actual appearance of the hypothetical aliens. Legends of serpent 'gods' abound through earthly myths and legends. The Canadian Museum of Natural Sciences created a humanoid entity with serpent-like features that was alleged to be biologically credible and was theorised to have been a possible evolutionary result from a dinosaur type of ancestor, had they survived. It was bipedal and humanoid with serpent-like features. Why would the creation story in Genesis prejudice its own credibility by having Adam and Eve produce two sons yet having responsibility for commencing the process of starting the human race. We have pondered this question in *Pillars of Fire* (Arena Publications). In the next chapter, we will analyse the 'cosmic connection' and ET possibly being an ancient and creative earthly visitor.

We have to ask, if ET is in our Earth space, how would they react if a different alien race became curious about us, due to our obvious signals and perhaps desperate for a new planetary home may suddenly begin to detect our intelligent radiations from Earth as they traverse the heavens, having been born in the confinement of the ships that had departed under the control of their ancestors long ago? Theirs, they hope, will be the generation to find the new planetary home and when they encounter the beautiful blue Earth, they are going to have it at any price. Fearsome weapons had been installed as part of the ship's inventory only to be used if all else fails. However, worlds such as the one their predecessors had left so long ago were known to be so rare in the galaxy, they must never pass one by, *even though it might be occupied!* These factors would produce a curious situation, alien against alien, with humans in the middle. This situation is unlikely but every eventuality must be considered.

These worrisome arguments win the day and the negative forces now prevail. A compromise is sought. We will become saints *and* aggressors. The negative genes are eliminated from true humans, but installed and nurtured in special 'android' constructions that are 'purpose built' for war, and are assembled in large numbers. They, and not the noble faultless humans, will do the defending and eliminating ....*wherever* necessary. Consider an army of 'terminators', practically indestructible androids, kept on standby for use when the occasion arises.

Traditional religious doctrines state that humanity was purposely created with an equal amount of positive and negative qualities, and the natural intelligence also bestowed would enable us to make the right and logical choice. However, those whose though processes deal only in logic might ask, what would be the purpose of such an exercise? Why not create all humanity as good, right thinking, well behaved individuals in the first place, and do away with the need for a 'hell' or place of confinement for those who fail the great test? It might be suggested, though, that there would be more risk of people

Dying of boredom in a world full of saints, than in one where negative traits also existed in humans as a challenge to be controlled and overcome.

Chapter IV

THE COSMIC CONNECTION

It may seem a little too convenient to offer a 'third alternative' to solve the mysteries of human creation in the form of visiting ET's encountering Earth and going about the business of creating and nurturing intelligence in a suitable subject but there are many people who subscribe to it and there is a wealth of circumstantial evidence to support it. In any case, with the rapid strides in the advancement of medical, biological and genetic sciences we may eventually fulfill the ancient Sumerian answer to the question why was man created. That is "To bear the burden of creation".   our world is just a tiny insignificant speck in the immensity and boundless infinity of the universe.   It is therefore impossible to think that there would be no other worlds containing even more advanced intelligent creative beings.   If not, as the late Professor Carl Sagan stated "It would be an awful waste of real estate". He also said "Something within us recognises the cosmos as home", but it seems that even among the members of the S.E.T.I. Team, some had misgivings, one scientist of the S.E.T.I. Team (Search for Extraterrestrial Intelligence) stated, "It is not just a case of telling Scotty to put the pedal to the metal", when contemplating the vast distances between the stars.

A logical statement, but isn't that just 'earthbound' thinking? 'we can't, so they can't? In any case members of his own science had postulated on the possibility of circumventing these vast distances by the use of Einstein/Rosen Bridges, and Wormholes in the fabric of space as theoretically possible and such considerations bridge the credibility gap when thinking of cosmic distances as being impossible to traverse. Our star system is so much younger than a host of others, yet in a comparatively short period of time, human endeavour progressed from the Wright Brothers to Mars Rovers in a hundred and some years.

Is it really so surprising to speculate that ET may already be here and clearly displaying this in the UFO phenomena? In addition, not as a recent event, as many believe in the possibility, that ET may have been seriously involved in our very existence, making the assumed ET presence responsible by inheritance, for the actions of their predecessors.

There is no question that the UFO phenomena exists, the question is are they artifacts from another world? Clearly, they have more than a passing interest in the Earth and humanity. To be sure, they would be aware that planets such as the Earth are a rarity in the cosmos and would be alarmed by the way humanity treats it. Ravaging the forests, polluting the seas, burying a horrifying bequest to our descendants and even polluting our Earth space with so much orbiting debris. We cannot fully control the nuclear beast we have created.

After the Chernobyl disaster, 6,000 people died and another 2.2 million people were affected by the fallout and there has been a 92% increase in childhood cancer. Great tracts of Belorussian land have been horrifyingly contaminated in the nuclear power station explosion that occurred in the 1980's highly contaminated graphite and uranium dust poured down on the land. If ET had been (including their forebears) in Earth space for so long they would have observed all these events but we still do not know for sure if they are here.

Have we finally convinced ourselves, largely through the use of the entertainment media and many written works and magazines, that extra terrestrial Intelligences *do*, most probably exist?

We seem somehow to have been subconsciously preparing ourselves for some kind of revelation, or is it the *they* have been preparing *us*? the idea is a few hundred years old. It has moved progressively forward from amusing science fiction to outright possibility on a par with our own scientific advancements and achievements. It is due to our technology that we are able to look again at events and happenings in history and see them in a different light altogether, and it is quite surprising just how far sighted some of the old science fiction writers where for their time, H G Wells, Isaac Asimov, Arthur C Clerk and others.

However, in spite of the previously mentioned mental preparation that we have subjected ourselves to, there is no good reason to suppose we will all take it calmly and without undue alarm. Did the ancient soothsayers and prophets really foresee such an event? The majority of the Quatrains written by Nostradamus refer to earthly events, wars and various dictators arising to commit their heinous deeds, and some have said to have been positively identified, such as Napoleon Bonaparte and Adolph Hitler, whose name was almost spelt outright. Nostradamus referred to him as 'Hister'. Nostradamus appears to have a special treat for us having a great 'King' coming from the sky. How could Nostradamus even envisage, regarding the technological advancement of his time, any king, or great leader, coming from the sky? one would have thought he would have hinted at some earthly source for his emergence, unless he had studied the biblical 'second coming'.

Although current humanity still regularly engages in war from time to time, it was a regular feature in the time of Nostradamus and sure enough "Mars (war) reigns supreme", with regard to the event reserved for us.

War and conflagration was much more prevalent in the time of Nostradamus but his prophesy regarding our century appears to have religious overtones. A second coming of Christ (The King of Kings) appearing with a host of angels in some kind of 'rapture' to overcome evil and separate the righteous from the wicked. However, it is entirely possible that it could be a second coming of an entirely difference kind.

Although we dealt with the Darwinian theory and its questions and anomalies in Chapter One, the cosmic connection is inextricably interlinked with it and in this regard we may refer back to it in this chapter from time to time, simply because the third alternative for human presence on Earth appears to clear up some of those anomalies.

It would be rare, I suspect, to find an anthropologist that would give any credence to the widespread assumption, or belief in many cases, that an extra terrestrial intelligence has been present on Earth for centuries and could have enhanced the intelligence and finer qualities in humankind. It would be viewed as professional suicide, but there must be many among their ranks that would admit that this theory is as good as their own at the present time and also, that it would clear up some of the puzzling anomalies such as the most un-evolutionary explosion of intellect, brain development and the finer qualities in the human.

Of course, even this theory has its critics for example, if the alleged ET's where so advanced in this science of genetics, how is it they did not get it right the first time and had to eliminate all the faulty examples of humanity in the great 'culling' programmer of Biblical Times. There are many references existing in Biblical teachings that almost prove the presence on Earth of other world beings interacting with certain humans we know as prophets or patriarchs. The same question (why didn't he get it right?) could be asked of a divine creator regarding faulty humans and indeed the chaotic events occurring in deep space, His other creation, which is far from perfect, but of course this question is neatly sidestepped by the ecclesiastics by teaching that it is all down to ourselves to survive in the afterlife or be damned, because we have been given the intellect to choose our own path, even to protect ourselves from cosmic threats. With regard to the ET creation theory, they could not be expected to be infallible and perfect and the mixing, blending and chemical effects within the cellular structure may be the most difficult of all the sciences. Having said all that, a mere human in the shape of Dr. Robert Plomin after a six-year search discovered a gene that was responsible for intelligence. This was shown in an ATV programme in 1997 called 'Equinox' on Channel 4. Since intelligence exists in other earthly creatures to a lesser and a greater degree, perhaps a better exercise would have been to carry out a similar intensive search for a gene (or genes) that are possibly responsible for the negative, evil and murderous behaviour evident in humans only, and not the animals. Nevertheless, it proves how human advancement is slowly catching up with that of our hypothetical ET creators, but it is doubtful that humans could ever pull off such a neat trick as providing a brain with two thirds of its cellular material 'in reserve' so to speak, that one assumes must be there for a purpose and should eventually be utilised. To refer gain to Dr Plomin and his team, since they were working on the discovery of the intelligent gene for six years and few of us then knew about it, perhaps they have been working on the discovery of the bad genes, as more than twenty years have gone by since their major discovery.

The Channel 4 programme *Equinox* had in fact highlighted the advanced capabilities of medical and biological science in a programme made a year earlier that the one quoted above, it dealt with an amazing head transplant operation carried out on a monkey, that involved delicate attachments of spinal cord and nerves and so forth and all the functions of the head where restored on the new body. The surgeon amazingly stated he was ready to carry out a similar operation on a human. Again, being over twenty years ago one wondered what has occurred in that time, after all many cadavers are donated to science. One with massive head injuries, yet a perfectly functioning body, perhaps switched off in a ventilator or could be matched with one where the body was virtually dead and incurable but the brain had all its faculties intact. The surgeon did say that he may go ahead anyway, despite ethical restraints. When considering the extra terrestrial supposition, only two valid assumptions are required, they are probably more advanced life forms on all those earlier formed worlds and the occupants could have reached Earth in a cosmic exploration programme long in our past. After all humans themselves are striving toward their goal of interstellar travel almost as though we where 'programmed' for it.

Add also the fact that we are fast becoming as efficient as our hypothetical ET 'creators' with such rapid advances in the genetic and biological sciences. Clearly, they are well versed in mathematics, or they could not have travelled to Earth in the first place, and or course, mathematics are part of the mental endowment we are bestowed with for such ventures.

Clearly, intelligence is a precious gift whichever way humanity came by it and a gift surely to be preserved and encouraged. People showing such gifts early in life should be encouraged, selected and 'streamed' to go further, because they are our future. They should not be held back by political ideology affecting the education services by such negative ideas, as 'level playing fields', and particularly (in modern parlance) be subjected to the 'dumbing down' process, lowering 'A' Level pass marks so effective that eventually even a halfwit can get one.

The strange fact is, that such a theoretical ET 'genetic' infusion programme, to create an over-endowed brain that may not have been initially very successful but finally was achieved, displayed in the Cro-Magnon peoples, it would explain the many anomalies in the theory (as per Darwin) of evolution from

Simian kind. The series of apparently unrelated hominid types appearing and disappearing, as though being purposely discarded, the paucity of bone fossil finds, failing to show a gradual evolution in any convincing manner. The obvious defiance of the creepingly slow evolutionary and natural selection process, with such rapid biological changes in muscle, bone and tissue, completely defying genetic stability that is so evident in primate evolution that is said to have taken some fifty million years to the present day primates.

We mentioned the Sumerian remark regarding the question why were humans created? Answer, "To bear the burden of creation". is that the whole plan? ET 'creators' arrive on Earth, select a suitable species to enhance their intelligence and creativity, who in turn will take up the baton and become interstellar 'creators' themselves. Why else would we be intent on mastering control of the human genome and advancing so rapidly in the biological sciences, and most of all for the compulsion for interstellar travel utilising the gift of mathematics, if it was it meant to be.

Once accepting this third alternative for the enigmatic human, it would explain the longevity of the UFO phenomena and the continuous alleged abductions of so many people around the world. If 'they' are looking for signs of human brain expansion in the positive mode, having given us all that excess or brain material, they would wish to see signs of a decline in the negative and retarding forces within the brain. So it would make sense that they are looking for 'uniformity' by abducting people from all over the world, which is exactly what is happening.

We must always return to the most important issue of the responsibility that such hypothetical creators would have inherited from their forebears, who decided "Let us create men in our image". Their equivalent to our Microsoft files would be replete with accounts of human behaviour patterns over the centuries. Sadly there would be as many accounts if not more displayed (than positive) in the negative mode. However, there are millions of people who live according to 'Gods Holy Law' or the laws of positive and civil behaviour. Quite unlike the dilemma, that Abraham had to endure, when pleading with the 'angels' (A.K.A) ET's to spare the lives of the righteous peoples in the cities of the plane i.e. Sodom and Gomorrah. The only problem was there were not any, which made it easy for ET culling programme to be implemented.

Again, when our mishandling of the dangerous forces of nuclear fission resulted in the Chernobyl disaster (an accident waiting to happen) and our actual utilisation of them in war, must have been a matter of grave concern to the ET observers. The doses of the contaminated graphite and uranium that rained down on the land was said to have been up to ninety times higher than that received by the victims of Hiroshima and generations to come will suffer its effects. Of course, the contamination that rose up into the atmosphere did not only sprinkle down on Russia, it was said to have travelled extensively and may have contaminated other lands, to a lesser degree.

An examination of meteorological records may confirm the following, the lady friend of an acquaintance of mine, had extensive flocks and cattle and shortly after the Chernobyl event, a group of boffin type individuals turned up on her farm with special passes of authority and scientific instruments and went among her animals, passing detection rods over the cattle. The lady, suspecting a connection to Chernobyl was told (when enquiring) "Just a precaution, nothing to worry about", and nothing else. However, she checked with other farms and they had the same visitors. It is quite amazing to think that such contamination could travel so far, yet one morning when getting into my car, I noticed it was coated with a thick orange dust, which was said to have travelled from Africa in the upper atmosphere and rained down on Britain. The point is, although our hypothetical 'creators' would have observed all these events, the intelligence to discover nuclear energy and its positive and negative effects was infused in the human brain in the first place by the ET's predecessors, so they have to bear some of the burden of indirect responsibility themselves.

In any case, prior to their superior advancement, surely, long in their past they may have experienced all these events themselves on their way to higher technology. All these factors would tie themselves to Earth, but of course other members of their race could be carrying out other profound creation programmes in the cosmos and they cannot just walk away from the results, wherever they may operate. We have said, that this extreme longevity regarding the alien presence, since before and during 'the aliens of Abraham', does not imply in any respect, that they mean us harm or that we should fear them, even though they did employ rather severe elimination' programmes in the past but by all accounts, the people on the receiving end, with regard to their behaviour patterns, seemed to deserve it. Of course, all of that only holds true if they actually did create humanity in 'their image'. It would be rather more disturbing to think, that they had nothing to do with our creation, and the fact that they had been here to long is that they have chosen Earth as their new home whether we like it or not. It would seem that most people of the Earth would prefer the 'alien creation' theory rather than a more frightening possibility. However, as said, ET creation would require a lengthy presence on Earth.

They would need to monitor their creations' behaviour patterns and activities all over the world for quite a lengthy period of their time. To do this they would need to operate from properly established bases. Furthermore, they would need to remove human entities from the Earth from time to time to examine their genetic material for signs of development into the beings that they envisaged in their hypothetical creation plan, and certainly much circumstantial evidence exists to validate *all* the above as possible. It does not require a great imagination to accept it, at least not to modern day humanity. We can also, obviously understand how the people of biblical times where so overawed by their presence in the form of 'angels' assuming this ET creation is true, it is clear that along with the positive creative and intelligent genes, the human brain still retains much savagery in its more negative proclivities, and the seemingly never ending series of wars and crime would be quite evident to any hypothetical extra terrestrial creator/observers, and would surely prompt some suggestion on a course of possible action by them as to what exactly to do about it, they certainly knew what to do about it during Abrahams time, but fortunately, for future generation, such genetic freaks could not pass on their faulty genes.

'They' may well be able to cure such human mental traits as easily as we might fine-tune a radio to get the best reception. The important questions would be, how would, or *could* they go about the task? If we assume that the floodings and the destruction of humans in those iniquitous cities in the Bible were not the acts of a divine God, were they the acts of creative 'gods' eliminating defective creations? More importantly, *might they consider it again?* Both divine creation and hypothetical extra terrestrial creation imply further action either an 'Armageddon', or by some other unknown means by hypothetical extra terrestrial 'creators'. But it must be true to say, that an extreme lessening of the type of behaviour exists in modern day humanity that was so prevalent in the alleged bestial debauching, that existed where the occupants of two cities in their entirety where infested with such sub-humans, where Abraham could not even find ten people trying to exist in the righteous mode among all those other people. Having said that, the debauchery of decadent pagan Rome a noticeable glitch in the assumption of the 'lessening' of such activity.

There is an obvious tendency in many people to only imagine aliens in the negative mode and only intent on subjugation of humanity. This is a far cry from wise all-knowing aliens arriving to enlighten us and bring forward our technology and is more akin to the type of scenarios depicted in science fiction movies of the fifties, such as *War of the Worlds*, a scenario that seems to be returning with, for example *Independence Day* a film depicting murderous aggressive extra terrestrial beings that were quite different with regard to the aliens in Stephen Spielberg's movie *Close Encounters of the Third Kind*, for example, or the lovable creature in his 'ET The Extra Terrestrial' that so pleased so many children, but almost all of the films made prior to them, especially in the fifties 'conditioned' us to expect the worst.

However, the logical view might seem to be that since humanity over the centuries, apart from recent astounding advancements, has behaved appallingly on the negative side and such behaviour is well represented in our entertainment media in aggressive action and war films, that if ET has been in earths pace for so long, any aggression by them in the extreme, could have been carried out long ago if that was what they wished and we would now all be subservient to them.

In the days of pagan Rome their evening entertainment was little different (to them) than our current visit to the cinema to watch a gratuitously violent film. The same gleam could be observed in the eyes of the watching audience as could, no doubt, have been observed in the eyes of the pagan Romans, only the dummies and studio blood have replaced the real things today, and a constant stream of violent and blood soaked films emerge from Hollywood, whether in the criminal or sci-fi mode.

We cannot predict the behaviour patterns of any off Earth Intelligence for obvious reasons, we can only judge and predict by our own standards, but it seems fairly logical to assume that by the time an earthly star ship slips into orbit around a life giving planet similar to Earth, we will have reached the stage of mental advancement to be able to respect any life forms found and not to alarm them unduly, but to study, analyse and observe, this assuming that we will (or have been allowed to) have advanced that far.

Of course, that would only apply if the object of the mission *was* for such a purpose. What if we viewed their world with covetous eyes? Since there is such an enormous amount of time left for human advancement and conditioning other worlds in our own system, and the alleged amount of life still left in our star, this scenario of humans looking for another planetary home can, for the moment, be discounted, but one day, as far as our descendants are concerned, will have to come to pass if there is to be any future for humankind.

Of course, if the universe is twenty billion years old, it has existed for around four or five times as long as the current age of our Sun. Therefore it is possible that other alien Intelligences could have evolved long before our Earth even formed, and so it is logical to conclude that the further future humans travel toward the centre of the galaxy the more likely it should be to encounter life.

Although the 'Fermi paradox' continuously asks "Where is everybody?" when considering the huge amount of alien Intelligences that *should* exist in 'the numbers game', we have to face the possibility that alien entities (assuming they are not already here) could encounter the Earth and position themselves in earthly orbit to gauge human reaction to their presence. Such a thing would be immediately classified above top secret and, if they did not make their presence known by taking over world communications networks we, the public, would only be enlightened when joint United Nations security decisions deemed it necessary. Dialogue, of course, we hope, would be attempted before any knee jerk reaction and panic by perhaps firing lasers at them. I dealt with this scenario in my book *Cultural Shock*. My conclusions where a bit worrisome given human behaviour when in the 'defence mode'.

Theoretically, such an event should not be such a traumatic shock for us, as said; surely, we have conditioned ourselves for it over decades with a constant stream of science fiction movies and data. However, they were watched between the popcorn and the ice cream, or from the comfort of our armchairs. Things would be quite different in reality. There *would* be immense panic, a stampede out of the cities to the hills, suicides, mass conversions to various religious persuasions, and so forth.

Governments are aware of this and would react in the aforesaid manner, in the hope that they could 'buy time' as it were, for some kind of programme to be set up to prepare the Earth for the cultural shock and social disorientation soon to descend upon us. We like to think we are quite civilised and in control but all that could quickly dissipate in such a scenario and our thin veneer of civilisation all be wiped away.

The reasons such hypothetical alien beings maybe in earthly orbit, could be quite varied. We could have invited them here by our escaping intelligent radiations, and since they are possessed with the ability to travel interstellar distances quite easily, they may have never encountered any other life forms and would be deeply curious and amazed by the proliferation of life on Earth. They may be non-aggressive and harmless, but once this became known to humans how would they fare against *us*? On the other hand, they *may* be desperate for a new home. Mineral depletion could have occurred even on their terraformed worlds and their sun may have begun to become dangerous to them, and perhaps excessive solar flares and outbursts of radiation all add up to departure or death. To have reached Earth it would be a one-way ticket for them, no going back, they must achieve acceptance one way or the other.

They may, after analyses of our world, notice the large uninhabited land mass of Australia, the Sahara and the Arabian Desert. They may consider a communicative request for living space on Earth and possess the ability to bring them to life quite rapidly. Worlds like Earth may be extremely rare in the galaxy. They are not likely to pass us by when we appear to have adequate living space to share. Governments all over the world have had over eighty years since the mayhem caused by Orson Wells and his fictitious 'Martian aliens have landed'. To formulate a plan on how to react to an alien arrival.

Others may come with a firm resolve to take the Earth for themselves and not be too particular how they accomplish it. Any alien race, if it was highly intelligent (and since they got here, in the first place they must be quite clever) would know after a prolonged analysis of human behaviour patterns, that it would be utterly impossible to live in peace with us. Humans cannot live in peace with each other, let alone an alien race.

They would be continually under threat of capture, attack and robbery by curio seekers and exposed to human undesirables of all types. They may think no more of eliminating us than we would spray a room full of midges. Any scenario is possible and this of course is the reason why formulating a plant to deal with it and prevent humanity from descending into chaotic destructive mode is so important. Of course, they need not lift an alien finger again us *themselves*. They would have noted our asteroid belt. They may have the ability to guide a couple of the larger ones our way, then just wait for the birds, who would be only creatures to survive, to clean up the mess, the waters to recede, and the weather to settle down and nature to 'readjust' itself just as it probably has had to do in the past after previous cometary or asteroid impacts that are evident from various eroded craters that have been observed by geophysical satellites. We can formulate many course of action 'they' may take but very few options we may adopt other than outright aggressive retaliation and possible extinction.

After the film *Rosswell* ended, a footnote added 'S.E.T.I has now been discontinued'. Frank Drake, who headed the programme, seems to have been given the task of preparing humanity for *something*. Have they found what they were looking for? How about this from Frank Drake himself, "I am telling my story because I see a pressing need to *prepare* thinking adults for the outcome of the present search activity, that is the imminent detection of signals from an extra terrestrial civilisation. This discovery, *which I fully expect to witness before the year 2,000,* will profoundly change the world". With regard to the remark after the above film, S.E.T.I is now operating again, perhaps as we have mentioned, to hide the fact that they have found what they were looking for.

Well, there it is. What could be clearer? Has it already happened? If not how do we become prepared? Of course, a signal from an alien source, although quite staggering and profound, would not cause us much alarm if it was from many light years away. We could feel quite comfortable about it, and not feel threatened by some race that was much nearer to us. However, like us, they would have to trust in providence that any planet encountering their signals will be peaceful, benign and not become a threat to them.

It is entirely possible that the length of time it would take an intelligent race to reach the ability to achieve interstellar travel, they would have mentally evolved as well, or perhaps intentionally bred out, negative genes from their race. Einstein said, "War is an illness of childhood". When we stop and think about it, it is quite true. It has been stated that war is a complete abdication from human rational behaviour. This of course is because as children we serve our 'apprenticeship' at it, playing war and war games. The sensible way, however (with regard to our communications), would surely have been to develop the cable system rather than allowing intentional radiations of an intelligent nature to leave the Earth uncontrolled, purposely revealing our position to other possible Intelligences, whose intentions may not be strictly honourable. We cannot guarantee the equation advancement – mental perfection. Our children are weaned on aggressive games wiping out alien star ships and every film out of Hollywood contains more and more violent aggression.

But, in any case, there is little point in worrying about it now when such radiations are now so far from Earth that many solar systems may have received them, and it is just possible that they may have developed *listening* apparatus but would not even consider transmitting, and maybe consider that the best plan is to quietly go and look over (assuming the ability exists to do so the senders of the signal).

For centuries, humans have been occupying their minds with thoughts of plurality of worlds' and possible means of travelling to them. First the Moon, seen as another world with possible life forms on it. Then our current advancement where our ambitions are much wider, methods of travelling to the stars are now seriously considered.

When we consider in noble thoughts, how we may bestow genetic advancement to help an intelligence to survive in our future space ventures, consider the old testament, it speaks of genetic implants to produce 'wonder men', such as Samson to slaughter other humans in battle. The entire Old Testament is ridden with questions and anomalies and one struggles to find 'divine' reasons for some of the questionable activity of entities accepted as, and called divine, 'angels', but they had, and did employ artificial insemination clearly described in the pages of the Old Testament when it suited their purpose to do so.

Although it would be preferable for many, in such a book as the Bible's Old Testament, to read only of divine holy acts, the plain fact is that many events have a quality that makes them very easy to view as possible interference and interaction by strange entities carrying out such deeds and actions for reasons best known to themselves, and for their *own* purposes, and it is very easy to view these 'angels', in regard to their activity, as extra terrestrial beings. Nevertheless, their primary interest is humans, for some valid reason.

Scientific advancement, which we seem powerless to resist, even if we wished to, is responsible for much doubt and uncertainty that gradually eroded into and refuted Old Testament writings. In the days of Moses, it was quite acceptable to believe God created the heavens, Earth and humanity in seven days, but eventually science suggested firstly that Earth could be millions of years old, then, of course, they arrived at that staggering figure of four and a half billion years, and if this was not enough, they added that men were not created by God at all, but descended from Apes, humans through the ages have been subjected to traumatic revelations of one form or another.

Most certainly, if one is comfortable with an attitude of blind faith, then one simply accepts the story without question. However, blind faith is not a natural feature of the human makeup. On the other hand, however, an enquiring questioning attitude surely is. Sadly, it has to be said, that the more science advances the more obscure and unbelievable certain biblical writings tend to become, but nothing science will ever come up with will ever negate the positive teachings of the New Testament and its efforts to promote and encourage the most acceptable behaviour in humans.

When looking around us at the more negative activities that humans are capable of, it could not be said that a divine, creative God would be entirely satisfied with his creations, that is, humanity and this quite naturally implies that he may wish to do something about it. Similarly of course, extra terrestrial biogenetic creators might also view the human mental processes as in need of some genetic correction and the constant and on-going removal of genetic material in all those alleged abductions, if true, strongly suggested a close monitoring process of human development over the decades. The assumption that ET's are seriously involved with humanity is always encouraged by such actions.

In the case of a divine God, it would be carried out in a 'second coming', or major event, where good positive thinking, law abiding, but not necessarily only religious people, would be separated somehow from the rest, where the righteous are saved and the rest are damned. This has been given a name i.e. 'The Rapture'.

Although it is accepted, that evil is a power, a force, few theologians now accept the concept of an eternal hell with no hope of appeal or reprieve when one is consigned to it. However, it could be asked can one believe in a heaven and not a hell? However, having said that, in reality they are both attitudes of mind and their existence cannot be proven one way or the other.

Another strange factor in terms of racial memory, is that certain humans behave as though they were striving themselves to create, or produce, special human beings and have done so brutally and under the protective banner of an evil, totalitarian regime, such as the atrocious medical experiments carried out by the Nazis and their attempts of creating racial purity, and a nation of blond blue eyed Nordic types. The same situation prevails in extreme racialism and ethnic cleansing and so forth. Nevertheless, the ability to create supreme examples of human beings may be nearer to reality than many imagine. Major advances in stem cell and genetic research may result in a situation where anything is possible and all of this serves to encourage an acceptance of ET creation of humanity as also quite possible and difficult to refute.

During a lesser and more unnoticeable degree, the educational process of 'streaming' and of course the '11 plus' grading in order to encourage youngsters toward a college education and then university was in place. For some time, naturally, this directed the more bright and talented and rejected the less bright in favour of the former. Of course, this did not sit well with the old Eastern Bloc dogma of 'everyone is equal'. Naturally, this did not include the stony-faced individuals lined up on the Kremlin balcony watching the military parades; they were more equal than the masses. Neither was it very acceptable to some of our own politicians, as previously said.

Now of course youngsters who have the talent can go straight to university if they have the ability to earn a degree without going through any 'grooming' process. Since the gift of intelligence has been installed in the human brain whatever way we imagine, it must always be encouraged and not held back by any 'level playing field' dogma from political ideology, pass marks should never be purposely lowered. The high intelligence bestowed on our Cro-Magnon ancestors must have served them well during the very significant event that happened twelve thousand or so years ago, geological disturbances, floods an extinction event that wiped out many grazing animals such as the mammoths. Whereas, their bones are found in great heaps. There is a noticeable absence of human remains mixed up among them. Nevertheless, it must have been an extremely traumatic event for our ancestors to deal with. Perhaps they where 'warned' of the disaster to come by means we can only guess at, and took the appropriate measures.

We have already studied, analysed and closely examined with our space probes most planets in the system. The initial driving force was, and still is, to find all the evidence of life off Earth. Our 'cousins' the chimps, are content and free of this driving compulsion to first wonder, then boldly explore until the only frontier left is deep space.

Humans have always been restless, first to know what was on the other side of the hill, then the other side of the Moon. Today, the search still continues. The intelligence levels in other earthly life forms are simply utilised to full advantage specifically and only in relation to their earthly habitat. A chimp would not give the Moon a second glance. A beaver could build a dam, but would not consider building a house. Other creatures, such as chimps, will *utilise* earthly objects like twigs and sticks but never create. Shellfish are cracked open by amphibians utilising stones on their tummies as they lay in their aquatic environment. A bird will utilise the force of gravity to drop a shellfish from on high to crack it open. Humans utilise the force of gravity to swing their space travelling created artifacts around the larger planets to propel them onward to others, and even outward toward the stars. It appears as though it is all precisely, what we are born (or created) for.

Humankind behaves as though they only have temporary residence on Earth. Of course, we know that Earth is the cradle of mankind but one does not live in the cradle forever. How do we reconcile firm religious convictions with this striving to leave the environment that we were instructed to 'subdue and replenish'. We have not entirely obeyed these instructions. One could suggest that mankind has subdued and 'depleted', rather than replenished. However, Earth's resources are a finite commodity and this depletion is a natural result of our efforts in technological advancement due to supreme intelligence and creativity. Gifts we did not seek, but which seem to have been bestowed upon us. If these special qualities had not been infused into the human we could still be existing like an ape colony.

The moment that a star switches on its nuclear, process it is doomed to extinction as surely as a human child is when it takes its first breath, everything that is created eventually expires.

The mysterious Tiahuanaco in the Andean Mountains was said to have been created by 'sky gods' in old Indian legends. They say that 'star gods' came in 'fiery chariots' and constructed the city. Well, somebody must have (and left without trace), there was simply no point in constructing a city in such a place *unless*, perhaps, the builders were used to thinner air.

Dr H S Bellamy studied the area intently and said, "The culture *behind* this culture is what takes our breath away" [46]. The city is said to be twelve to fourteen thousand years old. What is more, the much studied 'Gate of the Sun' is said to indicate, from the inscriptions, that this is the period that say the arrival of the Moon. Both Dr Bellamy and an Austrian engineer/cosmologist Hans Hoebiger, agreed on this indication. With all the aforementioned upheaval going on around the Earth, twelve thousand years ago, a world map of today probably differs greatly from what it may have looked like, and certainly much further back in time when continental drift had just begun.

With regard to the possible location of this divine creation centre (or extra terrestrial biogenic laboratory), Moses naturally has its location in his own familiar territory by mentioning the River Euphrates, but he also mentions Ethiopia and that one of the rivers, or heads, encompassed that land, which totally confuses the issue. However, its actual location may have been an area such as an island land mass, now sunk beneath the sea, or a flooded area, or even the Mediterranean, which has been dry in its history. Moses' account with regard to Eden is so vague that it seems clear he did not really know where it was located, but it has to be said that his writings appear to indicate that study of old historical documents which he would obviously have access to during his Egyptian upbringing in the Phaeronic household but not necessarily to later scholars due to destructive events carried out by the various pillagers and invaders.

Is this wondrous striving, creating intellect a gift or a curse? If it is a gift then the only hope we have of presenting is, passing it on, or bestowing it upon less endowed creatures, or allowing it to survive by our own proliferation, is to eventually leave the earthly environment that is doomed to extinction by the finite life of our star. *A thousand other suns are born every second.* We have a wide choice. Therefore, who, or what, process bequeathed the intellect to humankind ... The apes. It does not seem likely. A divine creative almighty God? ... Perhaps .... Extra terrestrial visitors in our past? ... Possible. It must now be clear to paleoanthropologist, and others entirely devoted to the study of past human ancestry, that at least a passing thought, as the years roll by with no solution in sight to the Darwinian concept, ought to be given for other considerations that might answer the questions regarding queries, anomalies and mysteries connected with an otherwise logical theory, and such queries and mysteries largely involve the enigma of *the human being.*

The striving in humanity to leave for the cosmos may be a subconscious desire to find our creator (or perhaps creators). Long before interstellar travel is achieved we will have conditioned and taken up residence on Mars and Venus and be happily depleting their still untapped resources. Our star was only half way through its life when humans appeared. We have plenty of time, but this force will never relent. There is nothing we can do about it.

Will we ever discover our creator (or creators) or will the creator, or creators appear or return to its creations in some kind of 'second coming'? Intelligent people who have strong religious convictions are to be commended simply because their intelligence has obviously prevented them from simply closing their mind and clinging doggedly with only blind faith, to their beliefs and tightly closing their eyes and hoping for the best. A 'second coming' should not be too traumatic for them, they have been taught to expect it.

Their intelligence has compelled them to think of other possibilities and their interpretations of life's wonderment and mysterious proliferation on Earth, compared to other hostile and hellish worlds, has made them conclude that some higher power was responsible for it all. However, the same intelligence must ask, being aware of the multitude of other worlds whether some have also intelligent created occupants.

They have survived the mental onslaught of science and its cold, impassionate values that refer to their beliefs as 'mystic and unscientific', and they have survived the science fiction data relentlessly pouring over us in films, documentaries, *Strange but True*, *The X Files*, the *Paranormal*, *Fortean TV*, *Star Trek*, scenarios, and so forth. They feel they can still maintain their beliefs. After all, the scientific view has so far failed to come up with the goods regarding the common ancestor they are all searching for, and we have not so much as a shirt button (that we have been told about) from an extra terrestrial visitor. The end result of all this, is that conjecture reigns supreme, and will do so until the answer is revealed one way or the other.

Many people who are struggling to maintain the comfortable, and surely most acceptable view, of divine creation, are now as vulnerable and susceptible to another alternative just as they surely were prior to the Darwin/Wallace revelations, simply because they cannot resist the relentless and compelling onslaught of scientific logic and alternative theories for their beginnings, constantly being put forth and discussed. They must feel they have been purposely led to a state of confusion by the conviction of people putting the Darwinian concept continually as factual.

In addition, no doubt, some feel let down, abandoned. They look around at all that is wrong in the world and try to reconcile it with biblical teachings of a return of the Creator to put things right, yet for two thousand years it has not happened, which brings us back to the point that surely one day, it must happen and this factor is embedded in the human brain and indeed was taught by Jesus himself. Logically therefore, those who subscribed to the possible ALIEN 'genetic creation' theory a similar conclusion must occur. An 'end game' if you will.

If one, therefore, can select a belief of their own choice and have faith in it, blind or otherwise, then this is surely more comfortable to the mind that constantly wondering and being confused. Those that are convinced that the common ape-like ancestor will one day reveal itself have chosen their path, and it may quite possibly come to pass. Those who have chosen to accept divine human creation have the most difficult tasks, and the more scientific discovery and revelation that occurs, the more difficult it becomes to maintain. However, they will always believe that there must be a higher power greater than man and that may be perfectly correct. However, even theologians today do not, as a whole, accept the basic writing of Genesis where the entire creation was wrapped up in seven days, or in the way creation of human kind was explained. However, how would people react to the revelation that human kind were created by the 'gods' rather than a single divine almighty God? Perfectly all right. It could be argued, after all, that God created *them* too, and they, in their turn appearing on Earth and accepted as 'angels', created m en in their 'image'. After all, the Bible used the plural, as though referring to a group rather than a single creator, and they are recorded as having said, "Let *us* 'make' men in *our* image". When one considers the many biblical deletions and alterations alleged to have occurred, if they seemed to be unacceptable to various emperors and kings and indeed by the Vatican itself, it is strange that the phrase "Let us create men in our image" was not deleted. After all, it took the task of human creation out of the hands of God the great creator and put it in the hands of a group, which the terms 'us' and 'our' obviously implies, yet it remains still in that form, surely a curious thing.

The whole theory of possible alien or extra terrestrial biogenetic creation is not a modern 'Star Trek' idea, or something out of the X Files. It was proposed decades ago. Even the late Charles Fort proposed that humankind might be 'property', in other words, an alien creation to be constantly observed and picked off the Earth from time to time to be checked, analysed and probed and others, such as NASA scientists, exobiologists and professors and other learned people, have considered this possibility. It seems indisputable that humans have a destiny in the cosmos unless our own negative behaviour patterns, or some cosmic threat that we have chosen not to prepare ourselves for, prevents it, we *will* eventually travel into deep interstellar space. Popular culture today, has it, that ET are actually working with top echelon scientists and the military to assist us, but for a price.

It almost seems as though we were pre-programmed from the start such an adventure. Why else would we have this amazing capacity for mathematics? After all, we never would have made it into orbit and achieved re-entry without such abilities to work out the precise manoeuvres and programmed them into our computers. Dr Max H Flindt said, "What applied to human space endeavours applies also to the extra terrestrials. If Earth *was* once visited by extra terrestrials, these visitors must have been well versed in mathematics. That is why I consider our capacity for mathematics to be an indication that we are *not only* of earthly origin".

That is rather a profound statement. Dr Flindt co-wrote the book, *Mankind, Child of the Stars*, together with Otto Binder the NASA scientist (Coronet), and they put up a very convincing case that humans are 'hybrid' creations due to biogenetic manipulations by extra terrestrial visitors in times past. In other words, our brains contain extra terrestrial genetic material deeply infused with intelligent genes that set us so far apart from our alleged 'cousin' the apes.

A few decades ago, such postulations may have been viewed totally as 100% science fiction. Today, when we are still barely out of the Iron Age, we see television programmes, such as *Equinox* shown on Channel 4, December 1996 that we mentioned on page 94 showing a head transplant or, if you like, a complete body transplant, involving monkeys. We heard the surgeon stating not only that he was sure it could be done with humans, but stating that he will probably *do it anyway*, regardless of any ethical objections. When I mentioned this earlier in the works, we must ask has it already happened and the 'patient' been sworn to silence from his gift of life and is being constantly monitored.

We hear announcements to the American Academy for the Advancement of Science by a certain Gail Naughton, that not only will human tissue be 'grown' in laboratories, but human organs, possibly even the heart, will also be 'grown' therein. More astoundingly, since that pronouncement the amazing '3D printing' may be utilised for this purpose we also mentioned, that in the following year. 'Equinox' was again enlightening us with the aforementioned Channel 4 programme mentioned on page 94 informing us in November 1997 that a certain Dr Robert Plomin and his team have discovered, after a six year search, a gene responsible for intelligence, and that others may well be found. Therefore, humans are living today through the greatest explosion of advancement and technology ever witnessed and may see the creation of life itself and even witness (ethical restraints not withstanding) infusion of intelligent genes into apes, which would no doubt greatly please all those ape watchers in the ape colonies, as they are doing their best to 'humanise' then as it is. This can only be the inevitable end to such research and experimentation, otherwise what is the point of it all in the first place? *Isolation of all those genes will lead to full control of them.* Learned people, such as Carl Sagan and others, have openly stated that Earth could have been visited in the past by extra terrestrial entities, and a large part of his life was taken up by exobiological questions and possibilities, being an original member of S.E.T.I and also involved in sending messages into space. Postulating what we ourselves *might* do in the future, we sense it has *already* been done in our own past.

When God said, 'Go forth and multiply', he did not qualify it by adding, 'in an earthly environment *only*, if you do not mind!' If we accepted divine creation, we would also be accepting that God gave us the means with our enhanced intellect in the first place that will allow us to do it. The same applies to those who suggest our geneticists are playing God by going too far in their experiments and future plans. Again, the ability to challenge the creator, or some might say 'creators', was given by them/him to do so in the first place. The strange fact is, that when we analyse all the factors for this startling possibility of alien biogenetic creation, they almost *compel* us to strongly consider it. In fact, these various factors enable the theory almost to 'prove' itself. Let us look at some of them.

We *know* sun-like stars exist and that some are not too far away, and that earth-like planets also exist. We therefore must ask, should extra terrestrial intelligence exist? The answer is that science *is* predisposed to the distinct possibility of their existence. Will such extra terrestrial civilisation *all* be less advanced than ourselves? Answer, not possible. Ours is a comparatively young star system. Extra terrestrial civilisations could have begun when dinosaur life forms were all that was evolving on Earth, and they covered one hundred and eighty million years of earthly dominance, and that is a huge amount of time for other creative intelligence to have evolved elsewhere in the vast universe.

If we accept far advanced extra terrestrial life forms, could they have once visited Earth? Answer, this possibility *is* accepted and may be reflected and help to explain the many depictions and legendary references to 'angels' and flying 'shields', 'chariots of fire', and so forth throughout history. Will *humans* ever advance towards the ability of passing on the gift of intelligence? Answer. We seem to have already started on the path by the aforementioned isolation of intelligent genes. Why can we not find bone fossil evidence of a clear evolutionary line from the true pongid apes to humans? Answer. Because they m ay not exist. The evidence, as it stand, appears more like a series of 'experiments' to enhance the intelligence of apes.

Will *humans* ever travel to the stars? Answer. Most likely. Will future human biologists, geneticists and medical scientists discover alien life forms and experiment with them? Answer. More than probable (long into the future).

Why does an extra terrestrial presence in Earth space seem to have been apparent for so long? Answer. A probable responsibility for our very beginnings.

Why do so many abductions continue to be reported where human genetic material is *constantly* removed. Answer. Possible analysis of human development over the generations, implying once again, a responsibility for human creation in the first place, and most of all an analysis of brain development toward the extinction of the negative and retarding genes in our makeup.

Who then, are these strange 'watchers' that have found their way into the earthly legends? In addition, why the universal racial memory of 'creation' by God, or gods, according to our own choice? So many claims of strange experiences and close encounters and alleged abductions involving strange unearthly beings have been made, it compels us to believe that these countless earthly reports indicate some kind of mass hallucination, or that there is some basis of fact in them.

The earlier suggestion, encouraged by the 'Condon Report' after a lengthy study carried out by scientists on behalf of the US Air Force into the UFO phenomena, that it is only a problem for the future physicists to explain, no longer holds good, there are too many other factors in the UFO phenomena to consider. However, the more obvious and high profile close encounters with apparently structured craft are lies, hallucinations or fact. These seem to be little middle ground. Are they all simply the aforementioned product of our subconscious wish to meet our creator (creators)? They exist throughout history, ever since we began to write it. In other words, as old as conscious humanity itself. Theoretically, this should tell us *something*. Perhaps they *are* simply part of the human psyche and always have been.

On the other hand, however, if this fantastic theory of biogenetic alien creation may have some basis in truth, then it may well be seen as quite natural for the 'creators' and their descendants, being who may (as humans, *also* one day) live for three of four hundred years to have spent an enormous amount of time in the vicinity of Earth.

If all these constantly reported objects in our skies and the alleged abduction claims, do signify an alien presence a constant monitoring of human communications and radio and television broadcasts would tell them all they needed to know about us. We have already been telling possible systems of the nearest sun-like stars all about ourselves for around ninety years or more. Our radiated communication emissions are now about ninety light years away from Earth and could be under the process of analysis right this minute.

Such entities that, we could assume, are currently in our midst, must conclude that humans are extremely slow witted, when they learn that their actual existence on Earth is continually denied by various authorities. After all the constant and, maybe, purposeful revelations of their presence, particularly with them having gone to the trouble of hauling great numbers of humans up into their craft to be analysed, poked, prodded and genetic material removed from their bodies and possibly inserting some form of homing devices to that they can come back and do it all over again *has* been alleged in such claims.

However, many senior people and *all* Governments, even though setting up investigating bodies, will continue to deny their existence. Apart from Governments who, of course, have their own particular reasons regarding their defence credibility for such denial, others will refute their (the extra terrestrials) existence subconsciously, perhaps because they strongly fear the possibility that it *might* be true.

It is impossible to deny the scientific factors that could enable it to be so. By that, I mean the necessary power sources for the feasibility of interstellar travel *could* have been developed by off-Earth Intelligence that simple factor cannot be logically denied. The lengthy period of time that such phenomena has been reported over the generations could well be an indicator that they *are* a natural phenomena. However, as said, that would be all well and good for the nocturnal 'Dancing Lights' and 'plasma balls' in landslip areas and ball lightning and so forth, but it does not address the problem of all the alleged 'close encounters of the third kind'.

Nevertheless, the general public has not been made aware of a single artifact that was 'not made on Earth'. If we wish to believe we will do so, and equally if we do not wish to accept it, then we will not. Even if a 'flying saucer' landed in Parliament Square we would assume, at first, that it was a stunt or a magician's illusion on a par with their ability to make jumbo jets disappear, and that they had simply reversed the technique.

Even if little green men emerged, we would assume that midgets in costumes were the answer. Unlike the hysterical days of the fifties, some at first, would probably laugh and clap and remark on how real it all appeared. There seems little chance, in any case, of such a scenario happening, simply because of they had the intelligence to get here, and then after observing our questionable behaviour patterns around the world, it would probably be the *last* thing they would consider. This assumed high intelligence would also allow them to be an unobtrusive as they wished, and keep us guessing for as long as they liked, which seems to be exactly what *is* happening. Clearly, with regard to one common factor in all the alleged abductions, they *do* wish (perhaps for the moment), to remain unobtrusive, otherwise they would not bother to apply the techniques of subduing and telepathically inducing a hypnotic command to the victims to forget it.

These extra terrestrial events we consider might be occurring in our own Earth space could very possibly be occurring in other worlds long into the future, with humans as the extra terrestrial. If we consider a few imaginative, but nevertheless, possible future hypothetical situations, it makes it quite acceptable that what *we* may subject another Intelligence to, may *now* be happening to us.

Our nearest star system (and therefore possible planetary systems), are situated only four and a half light years away. A triple system of stars, two fairly close, orbiting bodies, with a third further out sun, orbiting the other pair (Proxima Centauri). A ship, that had gradually built up a speed of only a quarter of the speed of light, would get there in sixteen years, much of which could be spent in a kind of 'hibernation'. If encountering a communicating intelligent life form, perhaps at the level of advancement of Earth two thousand years ago, it would be obvious that we would not immediately land and expect o be welcomed with open arms. We would remain as unobtrusive as possible and begin a thorough assessment of the level of their technology and, more importantly, a study of their behaviour patterns toward each other. Then, clandestine and furtive abductions would begin, employing any hypnotic or anaesthetising techniques, we may have developed in order to closely study the metabolism of the alien life forms. Prior to this, of course, a thorough study of atmosphere and planetary bacteriological microbes would have been undertaken. By that time, we would have become very efficient in manipulating DNA. However, once we began any tampering and manipulation of their genome, perhaps for what we may perceive as noble reasons, this would immediately cause a situation that we could not simply walk away from lengthy observations and study would be necessary in order to rectify any adverse effects in the species due to our manipulations therefore (just as we have speculated regarding ET on Earth) we would have brought a great responsibility upon ourselves and naturally, it would be necessary to remain attached to their world for some considerable time.

If the behaviour patterns of these hypothetical other world beings was anything like the way humans were behaving two thousand years ago, *and* down the centuries ever since, where, through all those centuries there was *always* a war going on somewhere on Earth, then any thoughts of contact would be immediately ruled out, and every operation would have to be carried out furtively and clandestinely.

It is interesting to note that in a hypnotic regression session of an actual victim of an alleged abduction, it emerged that telepathic communication had been made by the alien to the victim that the abductors felt that any open and firm contact with humanity would probably be detrimental to themselves.

Of course, any hypothetical 'watchers' in our own skies would not only observe negative traits in human behaviour patterns, they would assess the probability that humans, with the current rate of technological advancement, would most likely one day equal their own. Obviously, this would be a major cause of concern to them and would have to be dealt with eventually by their descendants. Of course, who is to say whether such possible extra terrestrial creatures on the early road toward their own ability to traverse space, did not display the same mental immaturity that humans display in constant and futile wars? Such things could equally have been a feature of their own past, perhaps thereby explaining this infinite patience with us, if such lengthy hypothetical observation of humanity *has* taken place. Therefore, any discovery in our own future travels of extra terrestrial life forms with negative traits, we would no doubt assist if possible but keep in mind our own murky past when dealing with them. This may also explain their tolerance and preference to evade, rather or destroy, the many earthly craft that have been scrambled to chase 'bogies' and alleged UFO's particularly in the more jumpy periods of the fifties and sixties, although today there is some evidence that we are firing lasers at them, photographic evidence of this exists in a piece of film taken from the space shuttle, yet they still do not retaliate.

In another work I pointed out that we do not have marauding hordes sweeping across the plains raping and pillaging anymore and there is more of a tendency today toward 'jaw jaw' rather that 'war, war', so the signs are that if we do advance toward the ability to penetrate interstellar space with power sources yet to be invented, that we may have suppressed, perhaps by genetic means, our immature and negative qualities altogether, and cease our incessant terminating of each other in war. Perhaps such suppression may not be necessary as the signs are that the human brain is slowly regulating itself and this would surely be noted by our hypothetical 'creators'.

If our own two planetary candidates, that is Venus and Mars, do not respond favourably to terraforming processes, perhaps we will be carrying out such activities on other more favourable worlds, light years away at some time in the future. Interestingly, the material said to be utilised in such terraforming processes is found in the most ancient rocks found on Earth, that is, over three billion years old, which of course, begs the question, whether such a process, long ago, could have been introduced to the primordial Earth? This substance would have needed a lengthy evolution of its own and we ask if earthly conditions prevailing at the time would have allowed this?

There is some evidence to suggest that life has arisen and reached a technological peak on Earth, then been utterly destroyed by cosmic causes. 'Ooparts', or out of place artifacts, have been found in coal seams or blasted out of quarry rock, millions of years old, indicating civilisation in antiquity. Earth could never have hoped to survive the commentary impacts that Jupiter suffered a few years ago. Most probably, all life would have been obliterated on the surface, and it would have been necessary, once again, perhaps for some sea creature to leave its habitat for dry land and possibly repeat the process once again.

One worrying fact is, that at least 90% of all life forms were destroyed on Earth two hundred and fifty million years ago, and since it takes our galaxy that long to complete one revolution, we are now back at the same point.

The major question always prevails with regard to the 'human question'. How it is that so many earthly legends have this one specific common denominator of their creation by God, or the 'gods'? Is there a link or connection with these alleged creation events and the equally alleged extra terrestrial presence in Earth space? An idea today, spreads rapidly due to international travel, which was obviously far less prevalent in ancient times, plus of course advanced electronic communication.

If human geneticists are making such bold postulations on what they will shortly achieve, here on Earth, what will they actually be achieving one thousand years from now? They will certainly be included among the crew of future star ships, along with all the representatives of the major sciences.

If we encountered a world with subhuman types of entities populating it, we might just decide to nobly infuse intelligent human genes into their makeup, to 'subdue and replenish' their world, and utilise its resources. Eventually these human activities would find their way into their legends and they were created by sky gods, and, of course, such gods would be ourselves, and some humans would remain and be seen coursing their skies.

With regard to these early legends and the cosmic connection, scientists will say 'not possible', or 'out of the question', and so forth and continue to insist that this massive over-endowment of intellect was due to a 'natural' evolution from apes which may well be so, but it seems amazing that allegedly intelligent scientists can so utterly close their minds to even the consideration of other possible reasons for the appearance of humanity on Earth, in favour of an unproven theory and that the sworn statements of abduction victims with supportive results from lie detectors test and many traumatic extra terrestrial close encounters mean nothing and that any consideration of a divine creative God as too mystic and unscientific, when they themselves have so many questions to answer and have not proved the theory (now a dogma) that they closely adhere to.

The odd fact is, that those people who choose to at least briefly consider the possibility of past Earth visitations, if not the occurrences of actual biogenetic creation, may be interested to see how many things, once unexplainable, now fall into the realm of the explainable, such as the following:-

1. Blue/green algae, said to be the substance earthly scientists will use to terraform other worlds found in Earth's ancient rocks. This complicated microorganism must have required a lengthy evolution itself. Could this have occurred on Earth over three billion years ago when Earth must have been a hellish place? What process introduced it into Earth's rocks, i.e. *over* three billion years old – was it 'seeded' into the Earth?
2. The sudden appearance of Earth's creatures in the Cambrian explosion of six million years ago, with no fossil history, yet possessing skeletal framework.
3. The proven 'Einsteinian' postulation on time dilation, where time virtually stands still for entities leaving Earth at light speed velocities, but time passing quickly on Earth.
4. The inability of our science to prove a natural evolution of humankind with a jerky procession of unrelated beings appearing and disappearing, rather like a genetic creation experimental programme under way.
5. Every earthly legend having a collective racial memory of men being 'made' by God or 'gods'.
6. The involvement of beings who ascended and descended on pillars of fire in front of the patriarchs, travelled about on moving stars and seemed so involved with humanity as to destroy them in large quantities with floods and bombs as malfunctioning mistakes of creation. Beings who were accepted as 'angels' yet behaved more like 'aliens'.
7. The long history of strange aerial craft and their occupants that are as old as writing itself.
8. The continuing and ongoing abductions as though our genetic development was being monitored. It has to be said, that this represents a considerable amount of circumstantial evidence, too much perhaps to be simply ignored.

One would have to be extremely unimaginative and closed-minded not to be the slightest bit interested in these logical conclusions and circumstantial evidence to at least consider the profound, but nevertheless possible, theory of *human genetic creations, by off-Earth Intelligence,* heavily reinforced by the almost immediate appearance of 'modern man' and the strange (possibly intended) retrogression of the strange Neanderthals. The fact is that. Oddly enough many features of the 'apes to men' theory support this 'third alternative'.

It certainly seems as though this mysterious enigmatic human brain is possessed with many racial memories enabling it to subconsciously recall and 'reinvent' things the possible donors of these highly intelligent genes did long ago. Does it seem a logical conclusion to dig brown powder out of the ground and expect to turn it into hard metal objects of iron and copper, and ultimately bronze, with time and aluminium closely following? Again, with reference to the odd 'ooparts', or out of place artifacts, aluminium items have been found long before our assumed ability to process it. The same applies to platinum items and enormous temperatures required with modern day apparatus. We could mention other oddities such as ancient batteries (proven to work, through experimentation) and evidence of 'electro plating'.

Part of this hypothetical infusion of memory and intelligence genes may be responsible for our incessant urge to travel to (or perhaps return) to the stars. After humans had achieved the ability to cause and control fire and the mining and fabrication of metals into artifacts, we were taking out first steps along the road toward the creation of star ships. Does it not seem rather coincidental that we should achieve these prerequisites so easily and also, somehow, to have been bestowed with the capacity for mathematical equation?

In other words, all the perquisites were neatly in place before the natural culmination of a space programme. These abilities, coupled with amazing inventiveness and creativity, ensure that future humanity, hopefully having eliminated naturally if not genetically, all the negative traits, will look on the galaxy, if not the universe, as their oyster. At that point, to find no evidence of any other life no matter how advanced would be incredible.

Four thousand years ago, humans were even then, when we consider the pyramids, creating on a mind-boggling scale, edifices that still astound us today. Although what we might term 'gifted' humans do exist today, particularly in the field of physics, searching almost for God in the tiniest particle of matter, genius 'per capita' seemed to be more prevalent in those times from the Sumerians to the ancient Greeks, who made many modern and scientifically sound statements, long before their time.

However, the great expansion of population seems to have diluted genius that not only occasionally manifests itself in notable beings, such as Da Vinci, Newton, Einstein and others. When the Greeks occupied Egypt they were less prone (unlike later conquerors) to destroy learned writings and may have obtained their wisdom from studies in the great ancient libraries indeed, *added* to them in Alexander the Great's time, by sending their emissaries all around the known world to collect works of wisdom from wherever possible.

It seems amazing that anthropologists can so easily dismiss these amazing human qualities out of hand and insist that we are simply 'naked apes', and that such qualities that seem more acceptable as a gift from God, or even the gods, are simply a simian bequest, however, if, as said above we do eventually travel deep into the cosmos and find no evidence of life, it would require a deep rethink regarding the purpose and creation of human life. After all, the majority of the planets we are discovering seem unlikely to sustain our type of lifeform.

This has been postulated when asking 'are we the first'? After all, as said, God may not have meant 'on Earth only' when telling the first humans to go forth and multiply. When we have reached that point in time, we would certainly know if ET created us or not. If not, one cannot imagine a more lonely feeling, but this would soon give way to a more proud and important notion that would make us see ourselves as 'gods'.

There are many other factors that seem to suggest that humans may be "not only be of earthly origin". In a work by the NASA scientists, Otto Binder and Dr Max Flindt, over two hundred pages are devoted to this distinct possibility, with sixteen chapters of 'clues', such as 'evolutionary clues', 'anatomical clues', 'brain clues, and so on, and there are many of them. If this book is available through the net, it is strongly recommended, see references.

Of course, the possibility exists that there could be an alien presence in Earth space that had nothing at all to do with human beings, and 'they' may be just as perplexed as those earthlings that take the trouble to review for themselves the paucity of human remains to study in a search for answers, assuming of course, that they take the trouble to research human origins, which would be difficult. Unless they could enter a major library late at night and mentally absorb all the current data.

It has been suggested that hypothetical extra terrestrial beings of high intelligence and wisdom would deem it unwise to expend resources on cosmic travel and the immense gulfs it would be necessary to traverse in order to reach other star systems, and would be more likely to utilise their resources on much closer to home endeavours relative to their future needs, when their own planetary resources begin to diminish, and thus concentrating on terraforming nearby worlds, or constructing huge orbiting environments, utilising their free solar energy, but we have no right to lecture on such things when we ourselves have ignored such advice ourselves when deciding on our own spending priorities.

Well, they may do all these things and more and, indeed, serious proposals have been made by learned people here on Earth for our own activities in this regard. However, the essential point is that where such logical conclusions as these are all well and good, they omit the possibility of such clever logical aliens being possessed of any of the qualities that ensure human adventure. Humans do things that require great courage and adventure and are done for challenge, or just 'seem a good idea at the time'.

It makes no sense to skydive, abseil down an office block, do a bungee jump, climb mountains, go down potholes and so forth, but the human qualities of adventurous spirit ensure that we do them anyway, and these same qualities will ensure that we cross our final frontier, not matter what the cost or whether it makes sense or not. We cannot fight this force in the human makeup, we are almost compelled to do such things, and if no ET's aliens exist in Earth space they eventually will do on another world, the ET's will be humans.

The strange situation exists that virtually says if extra terrestrial entities are not already in Earth space, then they 'ought' to be. This assumption is based on the 'Fermi' paradox, so named after the scientist who, like others, had sat down and did some real thinking and calculations on just how many other extra terrestrial civilisations there 'ought' to be out there. The number was quite significant, which no doubt assisted in their financial backing.

After the calculations of this group of scientists, astronomers and exobiologists who carried out these equations during their S.E.T.I or Search for Extra Terrestrial Intelligence programme, that utilised the radio telescopes of Greenbank, Virginia, Arecibo, Goldstone California, Nancay in France and Tidbinbilla, Australia, it came to light, after they had sat and held discourse on the amount of advanced civilisations there *ought* to be, that the question was asked, "Where *are* all these people?"

After lengthy calculations and computations of probabilities, it turned out that at least one, if not some of them, had achieved space travel and ought to have, by now, reached Earth or done so in our past. So again, they asked, "Where is everybody?", and this became known as the 'Fermi' paradox. Whether Professor Frank Drake, the leader of the S.E.T.I programme since its inception in the sixties knows something we do not, is not known, but he seems convinced that extra terrestrial contact *will* be made after the end of the millennium. In other words, *some time now*.

During the years of the S.E.T.I programme, one or two incidences occurred that made the scientists all sit up, one of which was the so-called 'Wow' signal. This seemingly extra terrestrial purposeful signal has never been satisfactorily explained (or repeated, for that matter), and caused the remark, 'Wow' to be annotated next to the data on the paper printout. It was definitely interpreted as extra terrestrial in nature which, in that context, meant it was not one of the previously false alarms that had raised a few eyebrows, but were eventually traced to an earthly source and had given the team some trouble on previous occasions. It would seem that this signal was interpreted as having the look of 'artificiality' about it. It was interpreted by Professor Jerry Ehman, who annotated the remark and it is still known as the 'Wow' signal today. As said, this sort of thing ensures the continuance of the programmed considering how profound the implications would be if it had continued.

The end result of all the aforesaid calculations was that there should be an enormous amount of advanced extra terrestrial civilisations, and the fact that many people surmise that we have not experienced a single event to suggest they have found Earth, or even actually exist (in spite of all the alleged close encounters, etc) probably reinforced the seemingly negative view that humanity may just be the first.

My own view, for what it is worth is, if we are going to allow that many planets similar to Earth exist by sheer weights of numbers, then they are not going to produce life in an orderly fashion 'one after the other', in other words, having some planet being 'the first'. Life is going to arrive spontaneously all over the galaxy, wherever these worlds are situated. Life on an 'Earth-like' world, if our own is anything to go by, is inevitable and prolific. We have to *fight* to keep it from swamping us. if the average householder totally neglected his garden, the house would eventually disappear under the foliage, as may be the case in South America, with many abandoned ancient places still undiscovered, that may increase the amount of strange links and apparent similarities with the culture separated by an Ocean and a Sea, i.e. that of the ancient Egyptians.

These strange factors seem to reinforce the legend of the lost land mass of Atlantis that stemmed from Egyptian legend itself. Other interesting factors support the legend, such as the 'Glomar Challenge', a ship equipped for core drilling finding evidence of *beach sand* (only forming on the surface) from drillings in mid-Atlantic. That area of the ocean known as the mid Atlantic ridge still quite volatile. Trans Atlantic cables have snapped with the ocean floor heaving up.

To return to the S.E.T. I programmed, a recent television film dealing with the well-known 'Roswell Incident' stated had now been terminated, it could not have been for long. It would seem a rather pointless and frustrating exercise to discover an obvious intelligent signal that, perhaps has not been intended for us, but passed over the world it *was* intended for, and it turned out to be two thousand light years away. It would take the same amount of time just to say 'we are receiving you', and two-way dialogue would be out of the question. No doubt the S.E.T.I programme has scanned Alpha Centauri and some of the sun-like stars that are fairly close, such as ten or twelve light years away, with negative results, but it takes a lot of time and there are many frequencies to scan, and at the rate at which they are being scanned another 'Wow' signal is long overdue.

With regard to Frank Drakes' prediction for the discovery of an extra terrestrial signal after the end of the millennium, the chances would surely have been prejudiced if S.E.T.I had closed down, but there are still plenty of radio telescopes in operation and an accidental signal could be received at any time, but we do not know for sure if the S.E.T.I team would be allowed to announce it without adequate preparation for its release. In 1947, the Roswell radio announcer reporting a crashed UFO was ordered to cease broadcast possibly by the FBI.

Frank Drake said in his book 'Is Anyone Out There?' "I do not believe the human brain is in any way limited. I think it can emulate the power of any intelligence we may find in the Universe *and I expect the discovery of extra terrestrial life to bear me out on this soon"*. That is quite a profound remark from a man who was actually *in charge* of the search. In his book, he all but says that we may have *already* made contact, and he seems to have been given the task of preparing us for the news, but as said, the authorities may have had a re-think on this.

As said, mankind's ultimate destiny with another star system is not only a vision or a dream, it is an absolute necessity if the gift of intelligence, in other words 'humanity', *is* to survive, not for a long time to come, of course, when the Sun begins to drift into old age and who know s what humans will look like in another four billion years, but the rather depressing situation with regard to human history exists, in that advancement seems to have risen and declined, as though some force decreed "Thus far and no further".

With the very brief amount of time, in cosmic terms, of our own ancestry we have missed all the geological and cosmic cataclysmic events that happened on Earth in the past, as evident in the eroded craters observed from geographical research satellites, and from what we know of the dinosaur's demise. However, these events could not have been too frequent or severe during their lifetime *before* the significant event of sixty-five million years ago, simply because we are told that their reign lasted for something like one hundred and eighty million years, this proving how rare serious elimination events are, the worrying point is we are a lot nearer the next one than we are far from the past event.

According to Frank Drake, "The Sun used to be thirty percent fainter than it is today". Apparently, it is the rocks of the Earth that provide this data. Now, if we turned down our central heating by thirty percent on a cold February night, we would notice it alright, yet the rocks tell science that the 'mean' temperature of the Earth hardly changed at all during this 'dimming' of the Sun. Furthermore, the rocks also seem to indicate that an average drop of only 5°C occurred during worldwide global temperatures, *during the Ice Age*. Once again, suspicion falls on the accuracy, fallibility and interpretation methods used in our checks, not to mention the design and correct application of instruments, and so forth.

So where is the thermostat that allows the Earth to self-regulate its bodily temperature? Is it the 'under floor central heating' of the magma chambers that supply all the volcanic out pouring?

Or, perhaps, the temperatures of the seas also warmed from time to time by the volcanic out pouring under the seas of the Pacific and the fissure running down the Atlantic, known as the Mid-Atlantic Ridge? This immense volume of magma must have some effect on the mean global temperature of the Earth.

We strip the Earth ourselves very firmly and rigorously of its Oxygen-giving foliage, and replace it with concrete *every day* on a massive scale, yet the same self-regulation seems to occur with the percentage of Oxygen, Nitrogen and other gases staying exactly the same. Humans on Earth could be viewed as a bacteria attacking a bodily cell irritating for a while but not destroying the body as a whole. Some scientists *have* described the Earth as comparable to a living entity in its own right. Logically, all the additional carbon , said to be emitted should affect the one hundred percent volume of the Earth's atmosphere said to be made up of seventy eight percent nitrogen, twenty one percent oxygen and one percent the other gasses, but no scientific announcement has been made telling us if these percentages have changed

If the Sun can dim in its outgoings radiation by thirty percent, let us hope that it cannot increase its output by thirty percent or we might fry. Perhaps these temperature drops were not the Sun changing at all but our solar system as it revolves along with the galactic arm passing through darker, dustier zones, which astronomers can observe in space, with lesser amounts of sunlight reaching Earth. The oldest earthly rocks are reputed to be over three billion years. Working on the basis of a billion being a thousand million, this means that with a galactic revolution of two hundred million years, the galaxy has swirled around twelve times within this period from the oldest rocks. The Earth revolving with the galaxy, dips in and out of the galactic plain, which may explain the periodic Ice Ages as the Earth moves in and out of the more dust-laden areas.

There are those who say we ought not to have spent a penny on S.E.T.I or any kind of space activity, for that matter, and that such research expenditure should have been used to address all the social problems that exist on the Earth. They have a valid point, but once again, these statements deny the existence of the very forces that are collectively called the human spirit.

We cannot stunt our own technological growth or quietly lock away each scientific breakthrough, discovery, and advancement, or suppress each achievement for some future age when we may be able to afford to exploit them better. Our brains would probably atrophy, like an unused muscle. The only sad point in all of this is that the first consideration regarding any new invention is whether it would be advantageous in warfare, indicating the present level of human mental maturity.

We cannot predict who will invent what and when and whether it arises at a suitable time for our balance of payments, and each scientific achievement since the utilisation of fire and the mining of metals, has been duly exploited in its turn, eventually leading up to the first blast off on the launching pad of humans into space, almost as though everything leading up to that point had been a pre-programmed part of the final result, but that was the more positive use of such hardware, where humans rather than nuclear devices occupied the upper portion of the missile.

The S.E.T.I programme was a brave example and a rare one, of science actually and actively, not only getting involved in the search, but actually postulating and calculating the possibility of extra terrestrial intelligence, when science in general appears either to ignore, deride, side-step or treat their involvement in the E.T.H., or extra terrestrial hypothesis as a form of professional suicide and to be shunned at all costs, and generally treated with the same attitude as the paranormal and other so-called fringe topics. Logically, one would think that with the alleged enquiring minds of scientists, it would be the other way around, nevertheless, the signs are, that more serious regard is now being applied to these anomalies and this is reflected in the abundance of TV programmes dealing with them.

However, the lengthy report and six hundred page book dealing with the Condon Report that wrote off the entire subject of UFOs and E.T.H. as 'Not worth investigating', proved this reticence in the part of science to get to grips with it. Dr Condon was selected, however, for his known views and cynical attitude to the problem (possibly intentionally) by the United States Air Force, with the special goal of damping down the hysteria of the fifties and sixties UFO 'flaps', and to get the United States Air Force, who could not adequately explain them, 'off the hook', so to speak.

A well know writer said, "In my correspondence, and in public, I debate with scientists. Our verbal battles are fair and honest, yet they are not objective. Paleontologists generally know too little of man's record, and what they have never heard of they are not interested in. Myths and old books are rejected as 'fairy tales'. They are terribly sure of themselves". It would seem that once their theories become documented and largely accepted they become 'set in stone', with a pronounced reluctance to abandon them.

The NASA scientist (a refreshing exception) Otto Binder, when referring to the work he co-wrote with Dr Max Flindt said "This book is concerned with the strong possibility, almost a probability in our measured opinion, that mankind on Earth may have had supper-intelligent ancestors from outer space and man, therefore, may be a 'hybrid', partly of terrestrial origin and partly extra terrestrial"… As this book appeared in the 1980's it is apparent that some form of genetic involvement with human evolution was being considered at least thirty or more years ago.

Even the simple consideration of what human scientific endeavours will undoubtedly achieve in the future, especially in genetics robotics and space achievements, fails to move some scientists out of their view, that only they and no other hypothetical extra terrestrial equivalents, could have done so long ago, and this is surely an enormously arrogant viewpoint, and such human advancements, therefore, must surely beg the question, *has* it all be done before? Nevertheless, many religious people still believe in an almighty creator and will say, "Even if ET creators where responsible God created them, so he is still indirectly our creator".

Although surely preferable, it must be said that divine human creation, at least in the form in which it is written in Genesis is scientifically questionable as, indeed, is the allegedly more scientific viewpoint of simian ancestry, and having said that, the theory, or possibility of extra terrestrial biogenic creation has at least equal footing as the former two theories for human origins.

If we assume that because of our currently advanced technology, and obvious future capabilities *do* suggest these advances occurred long ago on other worlds, we have to face the possibility that 'they' could be, if they emerged or began their evolution when earthly dinosaurs where evolving, a staggering two hundred and forty million years ahead of us. It surely does not seem possible that a civilisation could be that far advanced of another. There would be no possible way to know how to deal with them. They may have entirely disposed with a body form long ago and be just pure energy, or a 'life force', or 'spirit' – conjecture reigns supreme. The 'Fermi Paradox' once again arises, 'Where are all these people?" If all those billions of worlds are all barren and inhospitable what an utter waste of time for any divine creator it would, all be so utterly pointless to create such desolation.

However, as said, the possibility of an earthly presence of extra terrestrial entities, including the possibility of their involvement in human creation, is not only apparently easy to support, it is also difficult to argue against. It would be interesting to feed all the data into a super computer (if this has not already been done) and analyse the outputs. If it has been done and the readouts confirm ET human creation, there would be no question of a press release confirming it, and no amount of freedom of information would reveal it.

1. *We* have not discovered the means of interstellar travel yet, so *no* other civilisation that our computations tell us there should be a great multitude) have done so either – *illogical*.
2. *We* are the only intelligent race in the entire universe with billions of galaxies and billions of possible planets – *illogical*.

3. Every single witness of a 'close encounter' or the large amount of alleged abduction victims are all liars or mad – *illogical*.
4. Apes, who cannot even count their fingers, were indirectly responsible for the theory of relatively – *illogical*.
5. The fact that every single earthly legend contains human creation myths is just a coincidence – *illogical*.

It can be seen therefore, that we have an awful lot of seemingly logical assumptions to entirely refute or deny if we are not even prepared to consider the extra terrestrial hypothesis and their possible involvement with human beginnings as at least possible, so surely, the E.T.H. deserves at least scant attention by science. As for my own view, I believe it has received massive attention in Government, Military and scientific circles, they would not dare to ignore such a profound important issue.

It seems strange that geneticists do not seem to query, comment or *openly* discuss the comparisons of their *own* achievements in purposely bringing about genetic 'cloning' experiments in animals, and also in plants, which they know have been changed by an 'exterior force'. That is, *they themselves*, and, moreover, that the extremely slow evolutionary forces would not have done so alone. Then look at the aforementioned events on page 18 genetic changes in muscle, tissue and bone and brain cells in the rapidly changing human to produce our Cro-Magnon ancestors, and not consider it most unusual, when so obviously ignoring the normal processes.

Surely it must occur to them that an 'outside exterior force', currently unidentified, could equally have been responsible also to override and circumvent excruciatingly slow normal evolutionary processes. Geneticists must be very well aware of the genetic stability that is obvious in frogs, flies, fish, spiders and fern leaves, and so forth. When they review all these aforementioned rapid changes of all this rearranging and shifting about going on in the emergence of modern Homo Sapiens body form and skulls, it must at least raise the odd eyebrow among them, but they are all strangely silent (at least in public). Is it all part of this same fear of professional suicide and fear of being viewed as a nonconformist and slavishly adhering to the entire 'Darwinian package', as Richard Leakey would term it, at all costs? They cannot afford to ignore so many anomalies and questions forever.

Since their own skills can so obviously change so forcefully, could they not also concede that an unknown force could also have been at work on the emerging human?

Alfred R Wallace the co-producer of the 'Darwinian Package' said (largely in regard to human brain development), "That some intelligent power has guided or determined the development of man". If the great man himself can concede this possibility, why then cannot the geneticists themselves admit there *are* questions to be answered? The academic authority at the time may well have classified him as promoting the 'mystic and unscientific' viewpoint i.e. divine creation and therefore his views where ignored.

In spite of all that has been said about the possibilities of the E.T.H. this 'unseen power' that A R Wallace was referring to, was meant in the context of 'divine' forces, and he may be perfectly correct. No one on Earth is in a position to totally refute that the hand of God was not responsible in some way other than that set down in Genesis, which, after all, was written by a mortal man, who according to the British and Foreign Bible Society, was Moses himself. We have said that he must have been well tutored and therefore had access to many philosophical works, now destroyed and the term 'days' in the Genesis account many have referred to extremely long periods or eras.

The 'biogenetic' creation theory, though seen as perfectly possible and interesting, cannot be proven and must, for the moment, remain a theory just as the apes to men proposition must equally remain so, because each is as difficult to be sure about as the other, but where there is very little evidence at all for the apes to men theory, there is quite a large amount of circumstantial evidence for ET BC theory, with the theory of apes being any way responsible for it, coming as a rather poor third alternative.

It has been suggested, and strange earthly legends from India and from Pacific Islands tend to verify it, that humans came to Earth as some kind of 'cosmic refugees'. In theory, space should be littered with various star craft carrying desperate refugees away from volatile stars or other suns approaching their own star. Astronomers can detect colliding galaxies in space. Surely, within the two galaxies occupied with the colliding process must contain some planets with life forms advanced to the point of at least being able to make an attempt to evacuate and possibly live on elsewhere if they could find a suitable world.

It would not seem possible for two galaxies to collide without at least a few stars coming together with catastrophic results for any beings residing on their respective planets. Clearly, if such beings did attempt to escape the immediate threat, they would never be able to leave one galaxy and travel to another unless they were very seriously advanced, so they would have to carefully seek out as safe a haven as possible within the colliding pair. However, in our *own* galaxy, there have been exploding stars and if a reasonably advanced Intelligence could escape in time, they may get far enough away with living, procreating and dying on the ship until they reached a new home. In another work, I quoted a legend from a South Sea Island that sounded just like an account from distant cosmic refugees, floating through space in an 'ark' type craft, then finally reaching a new home and the landing being administrated by their onboard ships computer.

"In the beginning there was only empty space, neither darkness or light, neither land nor sea, neither sun nor sky. Untold ages went by, then the void began to move (possible orbital insertion) and turned into 'Po' (Earth). Strange new forces were at work (the effects of gravity). The night was transformed. The new matter was like sand that grew *upwards* (the descent) than 'Papa', the Earth Mother, revealed itself". (emerging from the ship to behold the Earth). With regard to the aforementioned colliding galaxies, astrophysicists have stated that a galaxy could move through another without any collisions due to the vast distance between the stars, a star only one light year away could not be seen as 'moving' but the above legend tends to speak for itself as a garbled account from someone attempting to record their arrival on Earth after knowing nothing but a cosmic existence. The author Robyn Collins also says that "An (Asian) Indian legend from the North West Frontier states that humanity descends from space beings who migrated to Earth and landed in the Lohit Valley". Perhaps a fleet of star craft, whose forebears had escaped that threat to their own world long before, landed in different earthly locations.

This scenario would also answer a lot of questions in evolutionary theory, and the advanced human brain having a lengthy evolution elsewhere in the galaxy. The story of Noah's Ark reflects strangely this survival of a select group escaping a catastrophic event, but occurring on Earth, again, it was written by Moses and may have cosmic origins.

There is a strange paradox in the similar earthly legends regarding the differently named 'Noah's all loading up their ships with supplies and attempting to save, and hopefully allow to proliferate, not only *their* life forms, but those of the other earthly creatures. Did the Noah's Ark legends have their roots in 'cosmic arks' packed with deep frozen embryos and life forms? Is this why animals seem to remain each unto their kind without change for so long into the past? We are not really in a position to answer any questions on human history earlier than 10,000 BC, and many perplexing questions asked by many authors decades ago on strange earthly artifacts, buildings, mysteries, legends and anomalies, still go totally unanswered. 'Catastrophe' seems to be a feature not only on Earth in its past but throughout the entire cosmos.

Returning once again to this amazing capacity the human brain has for mathematical equation, not once during the programming of all the computers on all the craft of the Gemini Mercury and subsequent Apollo programmes, right up to the tragic recent shuttle launch, was an astronaut ever lost due to faulty equations regarding the orbital insertion and re-entry manoeuvres necessary for space activities. This capacity for mathematics (that allows us to *have* space programmes) *never* failed them, at least in the Western world. The loss of life that occurred in the space programme was always down to mechanical human error and not faulty mathematical equations.

Regarding the instabilities of the cosmos, the threats and dangers to Earth, not only from cosmic debris, but ultimately from our star itself, only man has been pre-programmed with the ability to escape. All the other creatures of Earth are doomed to their fate, unless we choose to take their DNA patterns with us.

For those who view all life as a possible biological 'accident', is it possible that such an 'accident' would not only pre-place such qualities evident in the human, but also provide him with an escape clause or ability for a hasty departure to possibly survive and proliferate elsewhere in the cosmos if we discovered a suitable world, entirely due to the 'accidental' infusion of creative intelligence and mathematical capabilities?

When we look in our zoos at the obviously created and artificial environments of the reptile house and the alligators and crocodiles to ensure they do not die, then considering the apparent vulnerability of man with his hairless, flimsy body form, composed largely of water, yet possessing the ability unlike some creatures) to survive in *any* zone using the obvious intellect to protect himself, we realise just how adaptable the human body really is. Using this obvious intelligence for protection, we can reach out of the atmosphere, survive the conditions on the Moon, sink to the ocean's depths, and in the future, probably utilise this amazing ability to adapt and acclimatise to exist on other worlds far away from the Earth, until we were able to condition them ourselves to suit our metabolism, then dispose of our artificial survival units.

There are those who theorise that actual extra terrestrial entities would not at first appear in Earth space, but would be represented by analysing and information gathering probes, sent to possible life-giving stars in their local group. We might ourselves in the future, send probes to our 'local' group Proxima Centauri four light years away or Tau Cesti of Epsilon Eridani ten to twelve light years away, and these probes would cruise around any planets detected, analysing each in turn. Well this is perfectly reasonable. However, the sending of probes implies that they would be equipped with power source to enable them to reach their destinations quite quickly, and that the senders would not be so far away that they would all be dead before the information began streaming in from where they had dispatched them to in the first place. In addition, if the senders were that near, would they be able to resist going by utilising the same power sources to have a good look for *themselves?*

It is doubtful that if we had the ability to send a ship with a crew around the planets that we would have dispatched probes to do so. The emphasis is always on the search for *intelligent* life off-Earth, but we would be delighted, if not astounded, to find *any* life off-Earth, and if we encounter living creatures far into the future, we may have the ability to learn about their world by actually infusing intelligence into them to enable a kind of communication, or dialogue, with them. Just imagine conversing, for example with a dolphin. They seem to possess a very intricate and extensive communication system, and they have a long evolution, so, if we wished we could do this firstly on Earth and learn much from it.

Experiments with dolphins and attempts to learn 'dolphinese' have been carried out and may still be underway. The motivation to do so, however, was questionable, as it involves their exploitation in secret experiments in the hope, perhaps, that they could be used in sabotage attacks on enemy shipping, which of course is the human way regarding any new invention and its possibilities as a weapon.

When the aforementioned locating of a gene responsible for intelligence was discovered and its successful isolation announced to the world, how many of us knew that such a search was underway? What is going on in these remote desert locations, such as the well-known area 51? Even Presidents and Prime Ministers could, for all they know, be kept in the dark. Presidents and Prime Ministers come and go. Certain scientific research branches would not wish for some of their discoveries and highly classified activities to come out in some retired politician's memoirs.

It is known that certain experiments and research has been commissioned for certain universities in the USA and probably Britain, and certainly, in the USSR, into so-called paranormal activity, and fifty years ago at least, books were appearing dealing with 'psychic experiments behind the Iron Curtain, and so forth.

Certain anomalies of the human brain almost cross over into the paranormal, such as schizophrenia. No doubt, many cases that arose in the past were probably classified as 'possession'. The problem of schizophrenia is more widespread than most people generally imagine and the research and analysis going on in the attempts to properly understand it may also enable unexpected discoveries and side avenues to be made in the interesting field of delving into the workings of the human mind. Experiments seem to show that animals are not afflicted with this disturbing malady, only people and, although by the use of drugs we can tranquillize it, we cannot as yet cure it. However, the search for this cure could, as said, provide us with other unexpected revelations with regard to the intricate workings of the brain, which in general, we know so little about. The human brain can be manipulated; otherwise, no one would be susceptible to hypnotism. Over the centuries, certain people have learned to exploit their skills and power over others in this regard.

It is well known that the power of the mind is astounding and some people have learned to control and display it in various ways, from hypnotists to shamans and witch doctors in primitive societies. The power of the mind can literally kill or cure as Voodoo curses and 'miraculous cures' through strong faith or a will to believe have shown. However, many believe it is not the perpetrator that is responsible but the mind of the victims themselves that can affect the workings of the body.

There is an interesting case on record of two people called Charles and George, classified as idiot/geniuses, and although this appears to be a strange contradiction in terms, it seems to classify them quite adequately. They had been inmates of a Californian Institute since they were nine years old, yet they displayed one narrow field, sheer mathematical genius. No one had exploited them or tried to control them, but they manifested the strange intricacies of the human brain, which we cannot yet comprehend.

They knew instantly when any given date in any year ahead for twenty centuries fell on a Sunday. They would snap out without hesitation what day any date in the future will fall on, or produce the same data regarding the day any past date fell on. They would tell you the birth date of any notable being from history and how old they would be if alive today.

They would recall any day on their own lives and state whether it was cloudy, rainy or sunny and *never* miss. Yet were incapable of completing even the most basic sums of addition, subtraction or division. If asked how they do it they give a lopsided, drooling smile and say, "It's in my head". [54] I have mentioned this intriguing case in another work but it is certainly worth reviewing to highlight the strange workings of the human brain.

Imagine a person without the mental afflictions and also possessing these gifts of intellect as well. It is possible that such a genius could appear? Where did these enormous excesses come from? The human brain seems to be suitably equipped for even more excessive capabilities than it already possesses, unless they are signs that the fallow material in the brain finally appears to be stirring itself. If so, humans of the future may well be equal to the 'gods', mythical beings, who seem to have adequately impressed our ancestors in the past enough for them to record it all in legend. Everything seems to point to the stirring at last, of the mysterious over-endowment bestowed upon us at some point in the past of this excessive brain material beginning to manifest itself.

These strange traits that pop up from time to time in certain individuals may equally, therefore, be manifestations of great intellect genius that was more commonplace in the past and is now spread out with the enormous population increases into the worldwide gene pool. Certain factors may have affected or retarded to some degree, a natural mental development, such as the pollution and 'smog' of the industrial revolution, lead poisoning from past drinking vessels and in early motor cars.

The same source that related the story of the idiot/geniuses quotes a passage from the textbook on psychology, which states, "Arithmetical prodigies have been found among the ranks of the feeble minded. A favourite feat is to determine the number of minutes a person has lived from the date of their birth without paper or pencil, multiplication of three place numbers. Naming square roots and cube roots, or four place numbers, have also been executed within a few seconds.

Such people must have appeared in the past with perhaps, a poor background and to escape poverty with no university or study groups around to take an interest in them, or study them, they may have gone completely unnoticed, or perhaps felt they were odd and hid their talent for fear of being branded as a witch, or 'possessed', and perhaps wound up being exploited for profit for a mere bed and board in a travelling show, where people simply assumed their talent was mere trickery. An enormous amount of wasted talent in all fields must have arisen and faded away unnoticed, entirely due to the aforementioned fear of those possessing such qualities unless they appeared in the upper classes.

Oddly, science, except for one or two exceptions, shuns and ignores other strange manifestations connected with people and their experiences, simply because they cannot be scientifically tested and show an equally negative attitude to that metered out to their scientific predecessors, such as Bruno and Galileo, by suspicion and fear among the ecclesiastics. Clearly religious favour in the dark ages must have been very responsible for the aforesaid 'suppression if we ignore the sworn testimony of masses of people among them presidents, pilots, scientists and astronomers and assume also that the abductees *were* all dreaming (which might just be the case), and assume that no extra terrestrial manifestations exist in Earth space, then consider the 'Drake Equation', which was the result of the aforementioned study into the actual numbers of advanced technologies there *might* be 'out there', we could reflect on the possibility that *right now* inadvertent intelligent radiations from Earth now reach other worlds, up to ninety light years away, with many sun-like stars and possible Earth-like worlds orbiting them, therefore, now, an alien intelligence might be on its way, or at least *planning* to be soon on its way, toward the Earth. Hotel Earth is open – we have spare rooms – the lights are all lit up! In the twenties, when radio communication was in its infancy (and science fiction) had already postulated on the E.T.H., especially with writers like Jules Verne and H G Wells being so popular.

It seems strange that those involved in the development of radio, who knew that the radiated waves travelled at the speed of light, did not sit down and talk it over a bit. Largely due to H G Wells, most people thought Mars and Venus might contain intelligent life that possibly coveted the Earth for themselves. This makes the cavalier attitude with regard to the escaping radiations even stranger. They must have been aware that they would be responsible for announcing to other worlds that Earth contains intelligent life. Why did they not consider cable systems only? The answer, of course, is *expense*. This factor overrides everything, even placing orbiting protection (apart from the threat of military misuse or exploitation) around the Earth to protect us from cometary or asteroid impact. Naturally, it is important for our balance of payments to act responsible in this regard but as we have said, having the right people in place to assess the value of proposals put forward is vital.

In any case, we are today displaying this same cavalier attitude by *purposely* signaling to other systems and *purposely* putting descriptive gold-plated plaques telling possible extra terrestrial interceptors exactly where we are, and what we look like in our space probes, and have now sent them out of our system into deep space. This may seem to have been somewhat naive in expecting the recipients (if any) of such probes to be all peaceful with no negative qualities but it does reflect our major technical advancement.

Regarding, once again, the possible intervention of alien Intelligence in the evolution of the human, conservative scientists and geneticists will, for the most part, take the view it would be impossible, because they would not only judge by our own standards and state of progress, which may be thousands, or even millions, of years behind any theoretical extra terrestrials, they would say alien life forms and composition could be vastly different form our own. But would they? The human, or humanoid, configuration seems to be an ideal carriage for a creature with excess intelligence and creativity with upright posture, free hands and manual dexterity. Natural evolution in its creeping, snail pace way, may have been striving to produce such an ideal body form in the apes, and perhaps given another forty million years or so, may actually do so.

The odd factor connected with it seems to have been this 'exterior force' that A R Wallace mentioned, where a sudden boost, or guiding hand, seems apparent in the recent appearance of Homo Sapiens and fantastic brain development. With regard to the actual make up of the human body, its metabolism, physiology and so forth, most creatures of the Earth, in their wide diversity, have many basic things in common such as a skeleton, flesh tissue, blood, a brain, head, limbs for locomotion, eyes, nose, mouth, and so forth, all natural pre-requisites for an earthly environment but the human body form is built for higher pursuits.

If Earth-like worlds exist in orbit around those sun-like stars we have detected, especially if having liquid water the process in producing life there would very probably follow a similar pattern to that assumed on Earth. Why then should its creatures not have DNA and genes also? Almost everything in the universe is Hydrogen. We are aware of all the elements so far detected. All the atoms making up all worlds come from exploded star material. Why *should* things be so vastly different on other possible 'Earths'? If evolution generally is a force for betterment "Ever scrutinising, rejecting, and adding up all that is good" etc., then the human body form is a first class example of this, but it does not explain the excess of human attributes.

By the time future human geneticists have finished all their experiments with the huge range of creatures on Earth under the guise of attempting to produce better milk yields, and larger chickens and bigger cattle, and recreating the woolly Mammoths (now under way) from frozen tissue still being found, they will be very well versed in genetic studies and blends with many and varied life forms. Then they will be ready to board the star ships for the possibility of finding life on an Earth-like world (perhaps to be discovered) orbiting with Proxima Centauri, for example. Naturally, we are talking well into the future but it is comparatively easy to see where all this advancement is heading. Naturally, before this we will have utilised the knowledge to rectify all human deficiencies.

It really could have all been done before. This same finite amount of atoms and elements may also apply to bacteriological strains and nothing too terrible may be discovered in the atmosphere of other worlds, but the key word is 'adaption.

This will only have been achieved by the occupants, just as our common cold germs were so brilliantly used as an ultimate deterrent to war-mongering Martians by H G Walls. These things will obviously be among the top of the list in other world studies (should we ever encounter other Earth-like worlds). If this long suspected interest and concern with humanity *is* true, such entities will be aware of every bacteria and disease and could possibly cure every earthly malady at a stroke if only they dared bare themselves to contact and open interaction with humans, which appears from the testimony of certain alleged abductees, to be out of the question for the moment, entirely due to negative human behaviour patterns that we display on Earth, that seem to be reluctance to diminish along with obvious intelligent advancement and may have to be removed finally by genetic manipulation.

Their equivalent of a search for the Holy Grail may be to struggle, just as we do, to find a cure for all forms of cancer. This may never be possible as we will never be able to stop natural radiation from the cosmos. It is well known that the Sun would kill us all just as impassionately as it allows us life, were it not for the protective atmosphere and the ozone layer, which it is said is gradually diminishing due to human activity. However, we are fortunate to have a magnetic field that so efficiently disperses the solar wind and its dangerous radiation.

However, certain radiated particles know *no* barriers and stream through the Earth and ourselves continually. They have been detected by the photographing of the occasional explosive collision, with the atoms of the fluid underground 'bubble chambers' and various patterns, such as circle, whorls and spoke-like evidence of impact are shown. Surely then, as these particles stream constantly through the bubble chamber of the human body (70% liquid) they must also occasionally disrupt the atoms of human issue and produce the disorganisation in cells that cancer is said to be. These particles never stop, but even so, this is not to say that someday a solution to the problem will not be found.

Perhaps we will simply don a 'neutrino coat', rather like the hooded lightweight protection we use in a rain shower, but as said, we will probably never be able to stop some particles hitting the human metabolism and this will be a major problem for future astronauts on long space voyages.

There is a strange paradox in cosmic research and enquiry that frustratingly makes us aware that the more we advance, explore and discover the more we find just how little we actually *do* know. There are still a few aged people around today who have managed to retain all their faculties and with a little help can still get around a bit. Picture them looking down at a piece of Moon rock, then recalling the furor and excitement their grandparents witnessed as children with all the talk of the achievements of Wilbur and Orville Wright, and street corner newsboys shouting about it, a great achievement in its time.

At this rate of technological advancement, the people of Earth in many cases, must be more than a little disappointed that their limited lifespan will prevent them from experiencing the wonderful achievements still to come but they can take solace in the fact, that they have experienced in their generation, more cases of positive advancement than that achieved by any other generation. When holding their infant grandchild in their arms they would be bound to wonder what they would experience in their life to come. The greatest advances would probably be in the fields of biological, medical and genetic sciences that (we hope) should be concentrating on eliminating the savage, violent, criminal and regarding negative tendencies that are so frequently displayed in human behaviour patterns that so hold us back.

To return to the possible alien presence in earth space, even the amazing belief that many people hold that such creatures may even have been responsible for our very existence, their primary concern would surely be our rapid advancement toward leaving Earth to travel into their domain and whether considering the negative aspects of the human we are fit and ready to do so. However, this will not happen as quickly as we suppose many questions and problems are still to be resolved.

There are enough questions to keep the astrophysicists busy for many decades to come. This searching enquiring intellect we have been graciously bestowed with will eventually find all the answers. It would seem, given the aforementioned finite amount of elements and cosmic material and known worlds in the solar system, that a similar pattern will be repeated on countless systems throughout the universe, so given enough time we might eventually reach a point during the four and a half billion years left in our star, where we have advanced to the stage where we know *everything* there is to know. What then? Will our brains simply atrophy and wither or will we simply join the abode of the gods, become part of the circle of the creators, promoting nurturing and bestowing the gift of intelligence wherever possible on newly discovered worlds, perhaps in the same manner that we came by it ourselves. After all, stars like the sun had formed, lived their entire lives and died during the time our solar system was just forming, and perhaps being sprinkled with 'star dust'. Or in other words, the blue/green algae now being detected in the oldest rocks on Earth. Substance that, it is said, could be used to terraform other hostile worlds, such as Venus, into livable environments for humans which, it has been said could be done within a period of six hundred years.

One major concern, regarding factors that may disrupt our grandiose plans for the future, is the fact that there are many indications, that high standards of advancement have been reached in the past that were eliminated by either cosmic or earthly forces, and that humanity had to begin again as children, to learn and develop all over again. We may hope that we have not reached such a peak today. However, as long as more and more advancements are achieved, in being able to control violent earthly forces, and more importantly any cosmic threats from asteroids and comets, the more likely it is that we may avoid the troughs evident in past earthly history and remain at peak performance, being able to deal with any threat that presents itself.

Shortly before a group of dinosaurs, chewing their last meal on Earth, looked up to see another sun in the sky that seemed to be getting ever nearer and hotter and larger and scorching them out of existence, radio waves may have been wafting over the Earth from a source of intelligence similar to that of today on Earth. *Now,* they are *sixty-five million* years in advance of us. would anything be too difficult for such beings? We may speculate on previous attempts by ET to contact us in the past but other developed and advanced beings should in their turn, be still searching the cosmos we find that very easy to imagine other life forms on other worlds, but remembering that the dinosaurs ruled the Earth for one *hundred and eighty* million years and no intelligent life emerged at all in that time, finding a technologically advanced civilisation may be an enormous task. Moreover, only one planet out of nine in our system made it. Rocky, barren and large gaseous worlds may be all that would be encountered for millions of star systems in all directions. It is not surprising, therefore, that a space travelling Intelligence encountering Earth, which teems with life, are going to stay for some considerable time. It would take a very long time to study and analyse all the life forms that literally 'infest' the planet, Earth is bound to be a rare gem among all that derelict cosmic property. Nevertheless, in all that multitude, there may be worlds with humanoid creatures with basic intelligence and a workable lifestyle, much like the native Americans before conquest, who simply live at one with their world and hardly glance at the stars, let alone consider them as other suns. It is hoped that advanced human intelligence of the future would not attempt to impose, perhaps unwanted, advanced civilised culture upon them, whether they liked it or not, as happened on Earth in the past, and all but wiped out many cultures that hitherto had existed perfectly well without it. If we ever do answer all those questions about the Moon, perhaps we will find some answers about what happened to the assumed planet, now a heap of rubble that orbits between Mars and Jupiter. It may be possible to blow up a planet, but they do not explode of their own accord. In my book 'When the Moon Came', I put forward an interesting idea to consider with regard to lunar origins and its connection with the asteroid belt.

Clearly, there are lots of questions with very few answers. If the original material of the Sun spread out and threw off the material for the planets like astrophysicists suggest, why are not all the planets gaseous like those beyond Mars? Where did all the heavy rocky material come from to form Mercury, Venus, Earth, Mars and the assumed Planet X that every meteorite that streams into Earth's atmosphere was probably once part of? Surely the Sun would have kept all its heavy elements, simply because of its extra mass, to return to the special gifts humans have been bestowed with, including the amazing capabilities of the mind to cure or make the body ill or even terminate it, seemed to be hinted at in a strange experience related in a book that dealt with near death experiences. Many people have been pronounced clinically dead, stopped breathing and showed all the indications of death yet somehow come *back* to life later.

Out of body experiences are very strange and seem hard to refute, if told to doctors excitedly, as the patient found itself suddenly back in the body. They even say they moved to an area above the operating table and found themselves looking down at the team frantically attempting to revive them, and then floating through walls observing other activity in the hospital complex. In one case, a person in this 'un-dead' state, floated into the waiting room and observed a weeping relative holding a bunch of flowers and when again conscious described this to the doctor in detail, the clothes other person was wearing and how many others where there and even the colours of the waiting room décor. There are many surgeons who do accept out of the body experiences and life after death.

They can recall strange experiences that seem to run to a pattern, with common features, such as seeing and communicating with 'beings of light'.

Such 'beings of light' have often been associated with close encounter reports. In such a book, a chap called Mellon Thomas Benedict communicated with a being of light and it informed him that "Humans were given the power to heal themselves before the beginning of the world". Mr Benedict was clinically dead for one and a half hours, yet not only came back with all his faculties after such a period of Oxygen starvation should have caused brain damage, he also was cured of terminal cancer. Those things are startling enough, but the words of the being of light, if interpreted properly, back up the theory very strongly that the human brain contains genetic material, perhaps donated from 'beings of light', whose evolution goes back for many millions of years, and humans *are* now equipped to heal themselves, and in some cases, seem to actually do so at Lourdes. "There are more things in Heaven and Earth etc. etc.". The human brain is known to heavily influence ones health and well-being. Morose people constantly worrying about every ache will generally be less healthy than those with a cheerful and positive attitude.

It would seem that humans have a clear duty, if not to themselves, then at least to the hypothetical 'donors', to respect, nature, preserve and pass on the amazing gifts that lift us so far above the other animals of Earth, some of which compel us to do things we might simply do for no other reason than that the challenge exists. This is called the human spirit, yet strangely, there are those who misidentify it and reduce it to being motivated, or that it should *only* be motivated for reasons of monetary gain. A certain person, obviously totally devoid of these qualities himself, made the remark after the amazing human achievement for faster than sound travel on the ground. "What commercial value has it? We could not travel at the speed of sound around the M25".

This statement reduced this amazing human endeavour to the assumption that it should only occur with a view to making vast profit for the sponsors. Granted, commercial exploitation is a feature of technological advancement *after* invention, but the person who made the statement simply could not grasp the motivating qualities of the human spirit to do it in the first place, regardless of commercial value. This highlights the fact that there are many people who simply take for granted the advanced human abilities and do not dwell on the uniqueness of them.

Brilliant and gifted people seem to have been born out of their time in the past, when the lack of technological progress prevented exploitation and development of their ideas and inventions, which later proved to be viable. Leonardo Da Vinci is one example, and the even earlier alchemists in search of the philosophers' stone and their struggle to transmit base metals into noble metals, such as gold. Perhaps the philosophers 'stone' they were looking for was the knowledge of the atomic structure of matter, which even earlier Greek and Indian sages seemed to have had an awareness of. There are many such people mentioned in past earthly historical records and this indicates the 'peaks' that we referred to with regard to earlier advancement in all the sciences.

The ancient alchemists laid the foundations for future studies in chemistry where the word was taken from 'alchemy', also for the study of physics and astronomy. If the alchemists had possessed the ability to remove one proton and three neutrons from Mercury, they would have been able to transmute it into gold. The atomic numbers of mercury lead and gold are very consecutive when looking at the table of elements that had not even been compiled during the time of the ancient alchemists.

Modern studies in physics seem to suggest that future human scientist will possess the ability to rearrange the atomic structure of matter. Extremely low temperatures can retard the molecular motion of matter where individual atoms can be identified.

This apparent striving of earlier humans to achieve things impossible in their time, could be seen as another pointer to reinforcing the theory of some kind of racial memories within the brain, perhaps inherited from past achievements of the donors of intelligence to humanity in their memory genes.

There seems to be a strange trait in the human psyche of a 'death wish'. Some humans deliberately court extinction in various ways and it could be said that the entire human race is displaying such a trait by having the technology to do something about the threats from the cosmos, yet failing to do so. It must be said (to be fair) that verbal action at least is taking place with many clever individuals coming up with various suggestions on diverting cometary or asteriodal paths.

When Jupiter pulled a comet to pieces for coming so close to it, the various pieces came back and attacked it. Where Jupiter could take it and survive, the Earth could not and it would simply spell the end of life on Earth. What is the point of striving to achieve all this technology only to have it wiped out at a stroke? Our research on protective methods and operations should be proceeding at the same pace as other research and advancement. The 'near to Earth objects' are the biggest threat.

It would appear to be a waste of time to follow the course of simply changing their orbit as that orbit could be changed back again by a comet streaking in from the outer solar system in toward the Sun. the future astronauts will be mining them for substances in decline, perhaps on Earth. But the *more* we would destroy the *less* the threat becomes, especially the larger ones. One would assume we could atomise them with a nuclear probe getting close enough. Then if the shoal of rocks did reach our atmosphere, it would mostly if not completely burn up. If hydrogen bombs can obliterate twenty miles of built up area, destroying a five mile wide N.E.O should not be a problem.

With regard to worrying about *many* impacts instead of *one*, due to broken up pieces, conventional explosives might cause this risk, but surely an 'H' bomb, reputed to be able , as earlier said, to destroy everything out to a twenty mile radius, could take care of *some* of them. Obviously the largest one said to be one thousand kilometres in diameter, we would have little hope of destroying but such a body would probably maintain its orbit even if a comet collides directly into it. However, there are said to be around forty to one hundred thousand of them many with diameters of only a kilometre or two.

The comets that pass by Earth leave a trail of small particles during their periodic appearances, and when the Earth passes through the tail we experience meteor showers that look spectacular, but these pieces are very small. It is the asteroids that we should worry about. A recent amendment to a cometary threat seems to conclude that it will now miss, but originally a prediction had it colliding with Earth in 2,126. Some astrophysicists state that Earth *is* overdue for an impact event of a serious magnitude, the older fears of earthly panic and alarm may be prevalent, perhaps we are more advanced in our protection plan than people realise, but if announced an assumption would be that we are in danger.

We have proved that we have the technology to dispatch probes to accurately synchronise a course in order to closely encounter comets with photographic probes, such as Giotto, and our space hardware has even landed on an asteroid, so it can be seen that the technology is in place for a protective device to annihilate one that may become a threat to Earth. Giotto got *very* close the Halley's Comet but Rosetta actually landed on one. The Barringer Crater in Arizona is said to have been caused by an asteroid, or meteorite, about fifty meters only in width. Imagine the destructive force of an object kilometres wide instead of meters. The Earth's crust seems pretty solid to us, but so would an eggshell to a microbe. In my book *When the Moon Came* I dealt with this strange anomaly of crater depth to meteor size and the fact that the craters are so mysteriously shallow on the Moon.

The crust of the Earth is never more than forty miles thick. How could Earth survive an impact with an asteroid the size of a 'Ceres' for example, crashing into it? Ceres is seven hundred kilometres wide. Perhaps in the future we will devise the technology to change the course of the larger objects to collide with either the Sun or Jupiter, unless of course, the same attitude still prevails where costs, or political suspicion seems more prohibitive in relation to the severity of the threat, but the threat should never be underestimated. If, as we currently assume, that ET is among us, perhaps they could oblige us. After all, if they did create us they are not going to allow come pesky rock to undo all their creative efforts.

Looking much further ahead, humans may move to a much safer planetary environment than our own, where no comets, meteors or asteroids may exist. They may *all* be products of the destruction of Planet X, with the frozen mixture of planetary rubble, oceanic water and atmosphere making up some of the comets.

Although the possibility of mining the asteroids for certain minerals has long moved out of the realms of science fiction into scientific possibility, faster than light speeds, warp drives and 'worn holes' in space do seem still very much entrenched therein. However, an interesting theory to explain the problem of comets, asteroids and meteors, the Asteroid Belt itself, and even the destruction of an assumed planet and the termination of a once living environment with all that past existence of water on Mars, does exist and is dealt with in my book, *When the Moon Came.*

Although our cosmic knowledge has vastly increased since the days of Galileo, there are many more questions than answers, long ago (with the scientific data seeming to show fluctuations in the Sun in the study of rocks), conditions may have been quite different in our solar system long in the past earlier life of the Sun, which may have pulsations of very long periods that affected the proliferation of life in the outer planets, certain factors such as the Martian running water and life signs in meteorites (almost certainly from the Asteroid belt) are apparent. For all we know (or do not know) especially with the legends of human ancestors 'migrating from space', etc., life may well have originated in that region and move to Earth as conditions in the outer zone deteriorated. In things 'cosmic' we can now speculate on theories and possibilities that would have been viewed as sheer science fiction in the past, or even ridiculous.

As said, in comparison to the early astronomers, we know much, but many mysteries remain. No telescope in existence, even orbital telescopes, would ever see the dim reflected light of planets, but certain factors reinforce their presence, such as measurable perturbation influences, and also the following:- Our sun rotates once every seven days, and this is certainly not fast enough to have thrown off a ring of gas to form planets *now*, but an *earlier*, faster rotation could have done so, then ultimately forming planets, those planets would, in turn, have slowed the Sun's rotation to its present speed. Therefore, another sun-like star with a similar rotation may lend to suggest *it also* has a planetary group in attendance, but today we have more effective orbiting telescopes and other more advanced methods to detect other worlds and many have been found.

Stars are born, stars die and to be sure, all the atoms of our very bodies and those that make up everything we see around us, were once contained in a star. Since the same material permeates space and makes up the planets of other worlds, the amount of material constituents will be finite and perhaps certain universal laws make life *inevitable* rather than possible, and there is no good reason to suppose that such life will be un-fathomingly different in makeup. The fierce determination of life to proliferate is evident on any part of the world.

If we threw a handful of dried peas up in the air and let them fall on the living room carpet, they might eventually form the words 'well done', *if we did it a billion times*. if life molecules permeate space, as seems to be the case, then the old 'numbers game' comes into the picture to negate the 'life by chance' theory' and its relevance to the formation of suitable worlds for life to emerge upon.

Once the earth-like conditions are in place, the evolution of earth-like forms is easy enough to accept, and that the physiology of such life may have factors in common with earth-like forms for future geneticists to possibly blend, enhance, alter, as well as study, this brings us back neatly to the not too outrageous hypothesis that such a situation could possibly have occurred *here on Earth*, resulting in ourselves. It is as difficult to find a definite reason to refute it as it is to prove it, but the circumstantial evidence certainly has the 'edge'.

If all the alleged 'Rosswell Incident' and crash retrieval claims, with regard to alien craft and their occupants, *have* occurred, then we could quite feasibly have 'exobiologists', since the words means experts in alien forms. If *none* of these retrievals has occurred, then exobiology would be a science without a subject, with no examples of alien life to study, but there are certainly many earthly life forms for our hypothetical aliens to study.

The Rosswell Incident itself is now over seventy years old, so many exobiologists could now have amassed decades of experience. Since certain exobiologists, such as the late Dr Carl Sagan, were described as 'experts' in his field, it is interesting to postulate on how they *became* experts on alien life forms unless they *had* the life forms to study in the first place. Otherwise, clearly, they would only be experts in guesswork or the *possible* make up of alien life forms, but one can be an expert in postulation and imagination on alien life.

Certain scientists occasionally enjoy bringing the science fiction writers down to Earth with a bump unless the science fiction writers *themselves* were scientist as well, such as the late Professor Isaac Asimov, with a PhD in biochemistry and Professor at a school of medicines, it is not so easy to write them off as 'dreamers' with good imaginations and indeed some of their ideas have come to fruition. Arthur C Clark for example.

Certain scientist might point out to non-qualified writers, that this, that or the other could *not* be possible because of so and so, and so forth, but in doing this they are only pointing out how earthbound their *own* thinking is and how relevant it is to our own advancement.

If we consider again the hypothetical beings that may have begun their evolution on a far off world when the earthly dinosaurs were emerging then their capabilities would surely stretch the imagination of even the most gifted science fiction writer. Compare our own rapid burst of advancement from the Wright brothers to Mars rovers in one hundred and sixteen years. The mind boggles to consider what they may be capable of.

Therefore, when we think negatively and raise problems about the speeds, power sources necessary and dangers of exposure for excessive periods to solar radiation, micrometeorite damage, and so forth, we have to admit that all these *are* problems for *us* right *now*. However, to such as the aforementioned extra terrestrials, they may not be a problem at all. Therefore, the possibility exists that such beings could have been the far off visitors to Earth that have found their way into the earthly legends as 'gods', and whose descendants may well be *still* with us today. It would have been far easier of course, for them to proceed with their earthly pursuits and manipulations when being revered as 'angels' and not being subjected to any earthly threat from humans.

People do exist on Earth who are clever enough to have postulated on and produced interesting theories as to the propulsive forces and methods employed to enable the objects our ground radar installations have detected streaming into Earth space at speeds of four thousand miles per hour, then stopping dead over certain military installations, then abruptly moving off again and making rapid right angled turns, and so forth, completely defying gravitational and inertial effects. If these objects are controlled, structured craft, there is every possibility that if they can move at such speeds, quite effortlessly, in our atmosphere, then they most probably are capable of enormous speeds in the vacuum of space, but the faster one moves through space the more acute the problem becomes regarding impact with space rubble.

The threat of micrometeorite damage may be easily dealt with by some kind of energy field, radiated by and always ahead of the craft. In any case, away from the solar system there may not be any micrometeorite threat as every single grain of it may be entirely due to the destruction of Planet X, now a tumbling pile of rubble between Mars and Jupiter.

I have said before that it is totally incomprehensible that small, wispy gas clouds could muster enough gravitational effects to have matter accreting itself into solids from gas. They simply must have been part of a whole body and formed under immense planetary gravitational pressure, and that goes for every single speck of it, and most certainly, the huge chunks of solid iron placed in our museums, possibly parts of a planetary core.

Apart from another thesis on how comets may have been formed, a work on cosmic debris says, "Comets are something of a mystery themselves and astronomers admit that there is little certainty as to where comets originated, and current theories on the origin of comets place their source at distances so far from the Sun that the meteorites in our collection could not have formed there". This assumes that the meteorites we have all come from within a comet, but many arrive individually. In any case, some comets may have formed when the atmospheric gasses from the possibly destroyed Planet X, mixed up with planetary rubble, and may account for some of them.

The Sun continually burns off the icy material in continuing passes of the comets around it, where only the rocks remain and then perhaps spread out and become a meteor shower. One original supposition was that the continuous massive impacts on the Moon caused material to bounce off the Moon, with its comparatively weak gravitational escape velocity that would allow such material to depart and head for Earth. The analysis of the Moon rocks shows no 'pre-requisite' signs of life on them, such as amino acids but many in our collection do indicate this possibility for most of them. However, even quite recently it was suggested that we have a *Martian* meteorite, let alone a Lunar one. Great excitement ensued with what appeared to be a segmented worm-like construction but the excitement seems to have now abated.

Some of the carbonaceous chondrite meteorites hint at containing microbiological life signs within them. The secret is to collect them early enough and in zones where contamination by earthly bacteria is less likely.

Scientists collect meteorites from the frozen polar wastes for the aforesaid reasons. A recent UFO flap in the Chester area of Cheshire and the North Wales border, was said to be a 'short meteor shower'. This explanation will, no doubt, be cynically disregarded by UFO buffs as it has been used so frequently in explanations of the collective UFO phenomena. However, a certain evening newspaper on this event of June 1998, printed comments from an 'authority' on meteorites who stated, "It is estimated that every day *fourteen million* individual meteorites, mostly sand and gravel, enter our atmosphere and this amounts to approximately five tons per day".

If this was so, the sky would be permanently lit up with a celestial fireworks display, as it would amount to continual streaks every time we look up and most certainly a threat to orbiting satellites and the space station. When periodic meteor showers are expected, such as the Leonids, lengthy periods are spent looking up in expectation, when more UFO sightings (normally missed) would be reported.

The same article stated that "An American scientist had found a spoor of a fern leaf in a meteorite similar to the ferns that existed on Earth one hundred million years ago". This would have been a very profound discovery and would have been the subject of worldwide headlines on a par with the now uncertain Martian meteoritic life sings, and as such headlines did not seem to have appeared stating "Life discovered off Earth", the article must remain suspect, it would have been classified as an enormous discovery and conclusive proof of life elsewhere in the cosmos.

Moreover, if five tons of cosmic dust *are* distributed onto the Earth every day, then each year the weight of the Earth would increase by seven thousand tons. What would such continual weight increase due to the rotational speed and orbit? From the time of Romans, the Earth would already be enormously heavier. However, a more conservative estimate was made that much less celestial rubble enters our atmosphere every year.

Obviously, if we did have the aforesaid amount of five tons *a day*, then they would represent quite an important problem, because, even if they were only grains of sand it would be virtually impossible to maintain a space programme as every orbiting artifact, of which there are a great multitude, would all be peppered through with holes, and as said, the space station and of course, its occupants. Furthermore, a large amount of the incoming debris and dust would also, one supposes, be raining down on the Moon as it gets in the way of the flight path of the debris, but this did not seem to affect the Apollo programme or any of the many astronauts who visited the Moon.

Although contact has been lost with orbiting artifacts, it is usually thought to be of electrical failure and not attributed to micro meteors. The risk would prevail during the trips to the Moon and back and on the Moon itself, where consecutive astronauts spent a large amount of time, none of them were hit by, or seemed to experience, any impacts, and there were no frequent fluctuations of the sensitive instruments placed on the Moon to detect impacts as well as Moonquakes.

With regard to the hypothesis that every particle of this material was once part of a large planetary body, after Mars, the sizes of the planets *do* start to rise again and, if the supposed planet was midway between Mrs and Jupiter, it ought to have been at least the size of the Earth, possibly larger. It says in the book 'Cosmic Debris' "Mineralogically and chemically, they (the meteorites) resemble magmatic cumulates produced on Earth, which suggests their formation on larger parent bodies".

One assumption to be considered regarding the meteorites coming from a once whole planet would be that they all seem to indicate different ages. The planetary layers, going down to the core as they do on Earth, show different ages and, if they are, from a broken up world, then these indications would be quite natural and would naturally fluctuate.

Comets passing through the Asteroid Belt would disturb some and possibly set them on an orbit to eventually encounter Earth, but so also would collisions among the asteroids themselves, particularly with the mass of a large piece striking a less massive one. Planets do not naturally explode, but there could be other reasons for its destruction. One, straight out of a science fiction story might be a possible inter-planetary war. The attack, with enormously powerful devices, causing instability, and Jupiter's massive gravitational attraction, causing further out of balances stresses, eventually causing break up. The fearsome weapons spoken of the Indian Epic, the Mahabharata would certainly get the job done. They are written of so convincingly, it all seemed real.

However, it broke up, it is possible that its next-door neighbour, Mars, was on the receiving end of an impact with a mass of planetary rubble, which could have disturbed it greatly and brought about some kind of ecological catastrophe, making it unable any longer to support life. If a hypothetical Intelligence once occupied it, then they had the technology, they would ensure that at least a few escaped, but to where? …. Earth, perhaps?

It is said that some meteorites resemble Martian rocks, which have been analysed closely and even earlier, by comparison to the data analysed and computed from the Viking Lander soil samples carried out remotely on Mars in the seventies, then certain tests carried out on ground up meteorites could be used for comparison, and in any case, as said, Mars rovers are looking at Martian boulders all the time.

Of course, there was a rather sensational announcement of a possible primitive life sign found in what was suspected to be a Martian meteorite as mentioned on page 143. There would have to have been an enormous impact on Mars to eject planetary material, not only into orbit, but to escape the planet altogether, and for Mars to have possibly been impacted with a large portion of the material of another world, this might well produce the necessary conditions for such an escape of planetary material and send it toward the Earth.

To quote from *Cosmic Debris* again, Daniel M Barringer, who had a crater caused by meteoric impact in Arizona named after him, was said to have found traces of diamond in some of the fragments and speculated on its planetary origin, and the book says, "Estimates of the size of parent bodies changed considerably after 1960. Brown and Patterson, Wrey and Craig and Ringwood, followed Daley in considering that the planet from which they came was of lunar size or larger. They all assumed that considerable planetary pressure at the core was necessary to produce diamonds and other high pressure minerals". I have often referred to this possibility when continually hearing "They are material left over from the creation of the solar system".

Of course, carbonaceous material striking Earth at enormous speeds create the conditions of the necessary heat and pressure to produce diamonds, but they would surely be found loosely scattered, rather like the pear shaped tektites of melted and rapidly cooled material found from time to time and suspected to be from meteoric impacts, but as Barringer found his diamond traces inside a fragment that somehow survived atomisation, then having formed in the parent body is a logical assumption. We have said that the astronauts should have been collecting many diamond studded materials with the carbonation chondrite meteorites striking the Moon (if they did the data would be classified).

If the parent body, or planet, was hit by another celestial body of a mass near to its own, then its total destruction could surely have occurred. How could this happen? What happens to the outer most planets in a system where a star becomes a supernova and explodes, blowing off most of its mass into space? Could its supposedly reduced gravitational attraction cause it to lose its outer most world?

If the stars in binary and triple systems all have planets revolving around them, there must be some strange gravitational effects occurring on those worlds. However, with regard to the aforementioned supernova and its outer planetary bodies. Is it possible one could have its orbit disturbed and then become a free world? What about the theory of 'Nemesis', a supposed companion star to the Sun on which the film *When Worlds Collide* could have been based? If the Sun has a companion star as part of a binary system which comes by after a huge orbiting distance of millions of years, it may have carried an outer most companion along weakly until it encountered the pull of Jupiter that caused it to swing into the path of Planet X and collide head on with it.

Since every cosmic body in the universe tends to take on a spherical form in its initial formation, then it would seem logical that jagged irregular lumps of material represents the broken up components of an original, possibly circular, body. Some of the asteroids are very irregular in shape. 'Ceres', one of the larger portions is eight hundred kilometres wide, with 'Pallas' five hundred kilometres, and 'Vesta' four hundred kilometres. Many others are over one hundred and fifty kilometres in size, some with very strange shapes, which would make sense if they were all once part of a parent body and have been jostled and knocked about ever since.

Perhaps in future, we can secure a large amount of materials from them in mining processes and, more importantly, find ways to set the mined out hulks on course for the Sun, or Jupiter, perhaps even to Venus, to study processes regarding the atmospheric effects on the incoming body, the size before impact, and how much the atmosphere reduced it by future analysis of the crater, then terraforming processes making Venus a more livable place. In theory, there is nothing we will not be capable of in future space adventures, given the survival of our species.

Russian astrophysicists have calculated that Venus could be conditioned for human occupation in as little as six hundred years. The problems of the larger asteroids are possibly solvable in the future, but the question is, what do we do in the meantime? However, as said, even if the most unlikely direct hit by a comet did occur on the larger ones, it is doubtful that the flimsy mass of a comet would be sufficient to alter the inertial force of a five hundred mile chunk of rock tumbling along and make it change its course, and at the moment these very large asteroids are not a threat to us, unless disturbed.

It has been calculated that even to move a one kilometre wide asteroid an explosive force of six kilometres per second would be required. This of course refers to an immediate explosive force but suggestions are in place that would see a gradually imposed force over an extensive period such as directing sunlight onto its surface by a mirror for example.

To further support the theory that the large portions, at least, were all originally part of a whole (possibly) planetary body, M.I.T in the USA have analysed the spectra of them and *they were the same*. Moreover, carbon, an essential ingredient to earthly life forms, was found in them, so all the evidence for a once existing planet in the Asteroid Belt is in place. There appears to be plenty of circumstantial evidence to assume it to have been so, particularly the finding of carbon in its material, and reinforced by such factors as running water on Mars, implying that conditions for life seem to have been favourable in that part of the solar system. Then finding of familiar amino acids in meteorites and comments from scientist such as, 'Wherever this originated something once lived'. When adding up all the factors, there is strong circumstantial evidence for a planet to have existed between Mars and Jupiter.

Mars appeared to be our last hope initially, for finding evidence, however basic, for life elsewhere in our solar system, but the findings of the soil analysis programme carried out by the Viking Landers in the seventies were inconclusive, but little or no evidence of carbon was found by surface material analysis, at any rate, but now, in some areas, water ice has been found not far below the surface. The early Martian Landers only scooped and analysed surface dust, which leaves the question open regarding the makeup of the *lower* strata analysis, which will probably have to wait until we have set up a Martian laboratory on that world at some point in the not too distance future. Certainly, an awful lot of expense and scientific effort is being applied to that world.

Clearly, some quite profound 'shock event' occurred out in the orbit of Mars and the Asteroid Belt at an undetermined time in the past.

Everyone with at least a limited knowledge of physics knows that water boils quickly as pressure of the atmosphere is reduced, and would boil away altogether with the present conditions on Mars, or remain frozen at the Poles. When astrophysicists, therefore, make statements that Mars once had large quantities of running water on its surface, they are also saying it had a breathable atmosphere with adequate pressure and clearly temperate enough conditions to prevent that water from freezing up solid. In other words, they are saying life could have certainly existed there, simply because all those factors add up to an extremely earth-like world. Photosynthesis requires Co^2 water and sunlight, Mars has all of these.

Although it seems at first to be a staggering proposition, what do the following items of circumstantial evidence appear to add up to? Calculations indicating that the present amount of human brain material should have taken many millions of years to evolve having two thirds more than chimps, whose ancestry goes back forty to fifty million years. Legends of humans arriving from space in earthly mythology of Indian and South Pacific sources, temperate conditions appearing to have existed in the Mars/Planet X region of the solar system. Possible ecological destruction on Mars after the breakup of the assumed world between it and Jupiter and a population evacuation of some of its beings.

When we consider the conditions of Mars today, there could have been a time when living creatures (such as humans) could have stood watching pure water cascading along the river beds, breathed life-giving air, and looked up at a blue, sunny sky. If it was so Earth-like in the past, why not Earth-like creatures existing on its surface?

With the enormous volcanoes on Mars that, in the case of Olympus Mons at least, would dwarf Mount Everest, it is perfectly possible, particularly with all those dust storms that have been so severe that orbital surveyors have had to wait for them to clear to begin photographing the surface, that the surface dust hides the true makeup of that world, rather like the conditions around the area of Mount St Helena in the USA after its eruption, only in the case of Mars, those conditions exist all over its surface, the mount St Helena eruption buried many earthly dwellings.

So…. How far have we come, in our quest to solve the strange enigma of human origins and the cosmic connection that seems so interwoven in it? Strangely, the answer is, we have hardly moved at all. All the written data, all the reports of UFOs of all the different 'kinds', the masses of alleged human abduction and communicating with their occupants, the continuous analysis of human DNA removed from the victims, that should surely tell us something, in that they are continuous and not just a few examples to satisfy ET curiosity regarding humans. The apparent longevity of an ET presence on Earth back to the Biblical events, displaying what seems to be an advanced technology. Strange beings interacting with humans in such a manner that suggests that they are our masters and we are their 'property'. Many religious people keep their faith and in spite of the incredulous writings in Genesis are not dissuaded by the unproven Darwinian view of human origins and still believe in a Divine Creator to explain their origins. Others believe in a creator, or should we say 'creators' of the cosmic kind. Most people are not disrespectful enough to openly challenge religious people by saying "Show us proof of your God", although this was done in the book *The God Delusion* by Richard Dawkins, they simply take it or leave it. Religious people have every right to ask of anthropology "Shows us proof of your theory, we were once apes". As regards the cosmic connection, only circumstantial evidence can be produced. With regard to human development, the theory that we evolved naturally from apes is, at its best, tenuous, there just does not seem to be sufficient evidence, every time one 'expert' in the field claims to have found an obvious link, another will refute it. After all, there is intense rivalry in the field, all wish to be the one to claim the crown and go down in history for their efforts. That they may be pursuing a lost cause never enters their minds.

There is no real dilemma regarding Divine Creation, either you accept it or you don't. no circumstantial evidence exists that can be chewed over; it is no use quoting the wonders of universal creation in all those stars because there is so much chaos and destruction evident out there. On the other hand, both the Darwinian version and the extra terrestrial versions can be debated and are still today. One would suppose that one of the strongest factors regarding the ET genetic creation hypothesis is not only the wonders of the human brain (in the best examples) but the sheer longevity of what appears to the ET activity even pre-dating all those biblical events involving 'angels' in the Bible who were clearly written off as creating men in Genesis. That is, the ancient Hebrew writings in their old documents. One Hebrew legend states that King Nimrod's magicians spoke of a bright moving 'star' in the east swallowing four other stars. A 'mother ship' taking the observation craft back on board?

It seems that Abraham in his youth fell foul of the wrath of King Nimrod and was assigned to be burned at the stake, but a protective, being described as an 'angel' kept extinguishing the flames. For all that, the different opinions, arguments and rational discussion will prevail regarding the years to come or until such time, as a very traumatic event occurs to solve the issue once and for all. In any event, humans will continue their frantic pace of discovery and advancement that will ultimately take us perhaps from whence we came by our 'return to the stars'. A thousand suns are born every second, some will stabilise themselves into planetary systems not unlike our own. In four and a half billion years our Sun will be ready to die but those other embryonic Suns will be as old as ours is today, so, when the late Professor Carl Sagan said "Something within us recognises the cosmos as home", perhaps that is what it will all add up to, humans simply 'going home'.

We have spent a lot of time putting the case for the possibility of ET genetic creation and their lengthy sojourn on Earth, but both the religious factions and anthropology could say, "Show us just one piece of factual evidence that ETs had anything to do with human origins". Well.... Perhaps we can. Not in regard to any artifact dug up or discovered by erosion but within our bodily genetic makeup. The basic premise, regarding hypothetical ET 'manipulations' would obviously have to involve the infusion of intelligent advanced genes and those critical for the ideal human development. A recent TV programme dealing with the renaissance of the 'alien creation theory' that will not be silenced, mentioned a unique gene with no origins and not evident in other creatures appeared in the human genome some forty to fifty thousand years ago, during the great brain explosion that soon led to the onset of cultural evolution in the human. This special gene was named as the FOX P2 that was found in our nucleartides that set us apart from any other species and enabled communicative speech to develop. Was it purposely planted by ET?

Whether we accept the reason for human presence on Earth as a Divine Creational Event or that a supremely higher intelligence involved themselves in human development, is of little consequence, in the sense that it must culminate in a final revelation, which will, without doubt be a very traumatic event for humanity. What is important, is that the ecclesiastic hierarchy, the military and top echelon politicians of all countries meet together in order to formulate a plan to condition the masses for the obvious coming events, when all the conjecture and speculation must come to an end. It may be a 'second coming' in the religious sense, or, the final appearance and revelation to humanity of their extra terrestrial origins. It would be of great civil advantage to gently condition humanity that this event could occur and there would be no point in being frightened of it or reacting against it, because they would result in nothing but chaos, riotous behaviour and massive casualties for no reason. If the promised 'second coming' in the religious sense occurs, religious people are taught to accept it in any case. It need not include mass elimination of the evil and wicked, as they would whimper, kneel and repent in their masses, out of fear for their souls. In the extra terrestrial case, their involvement in human creation took place so long ago why on earth would they represent a threat? This conditioning by the Authorities should and must take place, to reinforce the view, that humans should react peacefully and embrace the 'new beginning'. Peace in the world, abandonment of futile conflagrations, a great technical and social advantages would occur and humanity would go forward as one in preparation for their membership of the cosmic brotherhood in our future space ventures. If this vocal conditioning of the masses did take place and was dealt with worldwide on every channel and was continuous, it would strongly prepare humanity to react in the current sense. All the indications are, that ET is in our airspace and has been for some time, therefore, this would in a sense solve their problems, as they may have been puzzling over how to carry out their second coming, simply because of human behaviour patterns. They would be well aware of the continuous attempts by the S.E.T.I team to contact their kind, so what better chance to reply through that medium to tell the world they come in peace. A scenario could develop just like that which was depicted in the interesting film *Close Encounters of the Third Kind*. A massive craft arrives at some agreed location and disgorges people, aircraft and their occupants, even ships and their crews. They would be younger than their descendants but would have some wonderful tales to tell. On March 5th 1955, Eugent Metcalf of Paris Illinois, reported a huge UFO, which he saw pursuing a jet aircraft. It caught up with it and seemed to engulf it, then it streaked away with the jet as captive. The same thing happened off the coast of Australia, which is well documented, where a young pilot and his light aircraft had a large object hover over him, which he stated, "it is not an aircraft", and he was engulfed by it. The hypothetical ET would know that it is the duty of Defence Forces to defend, therefore they have not destroyed any intercepting craft, sent up to investigate them, but many an earthly jet has tried to destroy them. This reinforces their peaceful intent. Therefore, this conditioning must take place. The message must be "Watch, learn, listen, have no fear, think only of the great advantages for humanity, whether it be God's final appearance or perhaps the 'gods'".

EPILOGUE

TOWARD THE INFINITE

When we contemplate the fantastic distances of the furthest galaxies in terms of billions of light years, we are looking back in time towards creation. Whether this creation was a spontaneous scientific event or a divine act remains for future humans to discover. The enormity of the universe and its cold, impassionate unawareness of human existence, and our petty worldly problems, staggers the imagination.

Here we stand on our rocky little world in the outer reaches of a rather unspectacular galaxy, twenty-three light years away from the centre, peering out to the most distant galaxies and wondering where they are heading at ever increasing speed as they retreat from us just as our galaxy is retreating from others. What chance will humans ever have of reaching them?

Even travelling at the speed of light still seems firmly fixed in the Star Trek zone, let alone 'time warps', and 'worm holes in space'. However, it is dangerous, with the pace of today's advancements, to state what we will never do and we might not all be aware that the data and dialogue used in Star Trek episodes is not the gobbledygook used in early sci-fi movies of the fifties, it is checked and approved by scientifically qualified people to ensure that the terms and processes used in the series are at least theoretically (if no currently) scientifically possible.

This particularly applies with regard to 'ANTI-MATTER' drive, which could be a force that may become controlled and utilised in the future.

If we consider once again, those remote fleeing galaxies, even if we could travel at just below the speed of light thereby avoiding the complications introduced by Einstein of the corresponding increase in mass, and so forth, it would take billions of years to reach the furthest of them and that would only be possible if they obliged us by standing still. However, galaxies do not stand still and it has been stated that the further away they are from us, the faster they are receding, possibly due to the force of the dark matter we know exists.

Very elderly people may remember tales from their grandparents who may remember the dire warnings regarding the probable inability to be able to stand up to speeds in excess of thirty miles per hour in the early days of motor car development and engine design, and the possible effects on the frail human body, people can only worry or reflect on the scientific advances of their day, when speculating on the future speeds of concord similar concerns ensued.

Today, if there was a four lane highway clear from John O'Groats to Lands End, a formula one racing car might make the trip flat out in around four or five hours. If it was possible to cover one hundred and eighty six times that distance in one second, then we would be travelling at somewhere around the speed of light. However, even travelling at this enormous velocity and maintaining it for millions of years, we would never reach the furthest galaxies, *unless*, of course, they began to slow down and perhaps 'met us half way', so to speak.

By this I mean that if the process, that is speculated upon by some astrophysicist regarding gravity, is correct and does eventually start to occur, then the gravitational attraction of the universe as a whole, assuming that it has sufficient matter dispersed within (that is not yet fully detectable) may start to have some kind of 'arresting' effect on the outermost galaxies and may cause them to begin to slow down a little and eventually perhaps not only come to a halt but start to move back to the initial point of departure of fifteen to twenty billion years ago, that is, the 'Big Bang' as it is commonly known.

We now begin a journey toward the infinite. Consider a futuristic scene. Eventually science advances to the point of actually utilising the aforementioned theoretical methods of covering huge distances by short-circuiting the obvious route, and our ship and crew are now out beyond the furthest fleeing galaxy and the limits of the universe. We slow our speed and look back toward the expanding brightness together with a huge spiral galaxy rushing toward us. Before we commenced our journey this remote galaxy was red shifted on our spectrographic analyzer as it sped away from us.

Now, everything we see is blue shifted as that once remote galaxy now rushes toward us. This little ship and its occupant's volunteers on a quest for our origins or eternity, hang in space like a mote, or speak of dust in St Pauls Cathedral. After all the clever scientific effort to get us to this unreal place, we do not wish for the approaching universe to simply drift over us and swallow us up again, so we contemplate on what is ahead in that pitch black foreboding vacuum, with not a particle or single atom of material or wispy veil of gas within it.

We power up our ship to maximum velocity and prepare to enter the sleep chambers, only to be automatically awakened by the ships brain if there is any good reason for it to do so. Can anyone imagine this dark, sub-zero vacuum going on forever without end? If you try to imagine eternity you will almost certainly fail. If one keeps up the thought processes of plunging on forever with no possibility of ever coming to an end, just continuing blackness, eventually

we have to abandon the thought process as the neuronal circuit begins to smoke a little.

If we could imagine the infinitesimal twinkling electronic spark of energy that is the thought, swiftly travelling along the neuronal path looking for a connection that would ensure a mental conclusion, it never makes one. It collides with the membrane. All that is beyond it is the film of liquid then the skull itself. Therefore, it bounces back again until it runs into the unused grey cellular material. No help here yet. It is not yet ready for the grand switching on ceremony, so back it goes again to continually bound back and forth until we give up the futile attempt at imagining eternity and allow the thought to come to rest. What would have happened if the thought could have made a connection in that spongy mass of unused neuronal cellular material? Will it *ever* be activated? Where *did* it come from?

The human brain is hugely over-endowed with unnecessary intellect as it is without this excess, rather like a legacy, but left to us by whom? Who *do* we thank for it? Will it, if or when it finally does become activated, allow that tiny, miniscule thought spark to make a connection with *any* question we ask let alone providing answers on eternity?

And so, the ship's computer, or on board 'brain', has powered up our ship to full velocity and switched off the engines. It has administered the freezing process for our bodyforms and constantly monitors the sleep chambers and the coal black emptiness ahead. It has no need to concern itself about micro meteors or dust particles or any damage to the hull of the ship. In the black sub-zero void outside there are no stars, no gas, no dust, in fact nothing at all, simply because not an atom of matter from the expanding universe has yet reached this area and may be never will, if the computed arresting process does begin to take place.

Long ago, the physicists placed a nuclear source in the ship with a fantastically long period of natural decay, and now it quietly provides the trickle of power steadily and reliably to supply the ship's brain, which in turn monitors our life support systems and feeds our metabolism from time to time with life giving vitamins.

The human body form back on Earth now looks quite different to that of the sleeping occupants of the ship's hibernaculums as they speed toward the infinite. Muscles and excess tissue have long dissipated due to lack of use and are therefore evolutionarily redundant. Our thin, spindly arms and legs are topped by a rather large skull where every brain cell within is actively utilised. We look slightly grotesque with this huge skull in relation to the slight body frame. This inordinately large head cannot now pass through the female birth canal and all births, restricted to a maximum of two, are carried by the use of Caesarean section.

All offspring are nurtured in retorts and parents occasionally take time out from their studies to condescendingly visit the rearing and tutorial colonies where machines 'programmed' the infants from childhood to adulthood.

Like ourselves today, who are told we have brute ancestors totally devoid of intellect and more akin to apes, their vast memory banks tell them of ancestors who started themselves off on the road to their present level of advancement, where three planets now have life supporting conditions, compared to the single planet whose resources are now seriously depleted and gave refuge to their ancestors.

These beings find it difficult to comprehend how things so clear to them were, according to their memory banks, so difficult for their ancestors to grasp.

The faint bleeping noises now became more pronounced. The sleeping occupants of the hibernaculums are beginning to show some signs of consciousness. The odd muscle will twitch, an eyelid will flicker, they are gradually becoming aware again of their conscious selves. No one wants to open their eyes. Even through their eyelids the bright searing light makes them screw up their eyes and fear to open them. In a few seconds, arm muscles will allow them hopefully, to raise their hands to their eyes to shield them from this amazing light everywhere ……. Yet now, the light seems warm and inviting they seek advice from the 'Brain' … "My response is, you must go into the light".

END

REFERENCES

CHAPTER I

1. Time Life International – (Evolution)

2. Johnson and Edgar (Wiedenfeld & Nicholson) 1971

3. Mankind Child of the Stars? (Coronet)

4. From Lucy to Language (Wiedenfeld and Nicholson) 1996

5. The Story of Archaeology (Wiedenfeld and Nicholson) 1996

6. The Origin of Humankind (Phoenix) 1994

7. From Lucy to Language (Wiedenfeld and Nicholson) 1996

8. From Lucy to Language (Wiedenfeld and Nicholson) 1996

9. Mankind Child of the Stars? (Cornet)

10. Mankind Child of the Stars? (Cornet)

11. From Lucy to Language (Wiedenfeld and Nicholson) 1996

12. From Lucy to Language (Wiedenfeld and Nicholson) 1996

13. Time Life Nature Library (Evolution)

14. From Lucy to Language (Wiedenfeld and Nicholson) 1996

15. From Lucy to Language (Wiedenfeld and Nicholson) 1996

16. From Lucy to Language (Wiedenfeld and Nicholson) 1996

17. Subdue the Earth (Granada Publishing)

18. Cosmic Debris (John A Burke)

19. The Neanderthals (Jonathan Cape)

20. Making Silent Stones Peak (Wiedenfeld and Nicholson)

21. Making Silent Stones Peak (Wiedenfeld and Nicholson)

22. TV (Horizon 25th September 1997

23. Mankind Child of The Stars (Coronet)

24. Mankind Child of The Stars (Coronet)

25. Mankind Child of The Stars (Coronet)

26. The Origin of Humankind (Phoenix)1996

27. The Origin of Humankind (Phoenix) 1996

28. The Origin of Humankind (Phoenix) 1996

29. The Origin of Humankind (Phoenix) 1996

30. From Lucy to Language (Wiedenfeld and Nicholson)

31. The Origin of Humankind (Phoenix) 1996

32. From Lucy to Language (Wiedenfeld and Nicholson)

33. The Origin of Humankind (Phoenix) 1994

34. The Origin of Humankind (Phoenix) 1994

35. The Origin of Humankind (Phoenix) 1994

36. The Origin of Humankind (Phoenix) 1994

37. The Origin of Humankind (Phoenix) 1994

38. The Origin of Humankind (Phoenix) 1994

39. Children of the Universe (Futura)

40. Mankind Child of the Stars? (Cornet)

41. Mankind Child of the Stars? (Cornet)

42. Darwin Century (Double Day)

43. Darwin Century (Double Day)

44. Great Events of Biblical Times (Wiedenfeld and Nicholson)

45. Is Anyone Out there? (Simon and Schuster)

46. Mankind Child of the Stars? (Cornet)

47. Mankind Child of the Stars? (Cornet)

48. Did Spacemen Colonise Earth? (Mayflower)

49. Mankind Child of the Stars? (Cornet)

50. The Near Death Experience (Routledge)

51. Cosmic Debris (John C Burke)

CHAPTER 11

None

CHAPTER 111

None

CHAPTER IV

None

www.ingramcontent.com/pod-product-compliance
Lightning Source LLC
Chambersburg PA
CBHW020804160426
43192CB00006B/432